Photographic acknowledgments

All illustrations in this book have been taken from
out-of-copyright material held in the British Library's
collections, with the exception of those which carry a credit
line in the accompanying caption. Patents are numbered and
dated for ease of reference.

OXFORD UNIVERSITY PRESS

Oxford   New York   Toronto
Delhi   Bombay   Calcutta   Madras   Karachi
Kuala Lumpur   Singapore   Hong Kong   Tokyo
Nairobi   Dar es Salaam   Cape Town
Melbourne   Auckland   Madrid
and associated companies in
Berlin   Ibadan

Library of Congress Cataloging in Publication Data

Dale, Rodney, 1933-
   Timekeeping / Rodney Dale.
      p.64, 24.6 x 18.9cm. — (Discoveries and inventions)
      Includes bibliographical references (p. 63) and index.
      Summary: Describes the changing ways of thinking about
time, the emergence of the calendar, and the quest for a more
accurate means of measuring time.
      ISBN 0-19-520968-0 (acid-free paper)
      1. Time—Juvenile literature. 2. Clocks and
watches—History—Juvenile literature. [1. Time. 2. Clocks
and watches.]
I. Title. II. Series.
QB209.5.D35 1992
529—dc20   92-21661
                                                            CIP
                                                            AC

ISBN 0-19-520972-9 (paperback)
ISBN 0-19-520968-0 (hardback)
Printing (last digit)   9 8 7 6 5 4 3 2 1

Designed and set on Ventura in Palatino by Roger Davies
Printed in Singapore

Front cover
**John Harrison's first sea clock, known
as H1.**
Natioanl Maritime Museum

Facing page 1
**Calendar page for March from the
Bedford Hours c1420.**

Page 1
**Denison's four-legged gravity
escapement.**

Pages 2 & 3
**Detail from Hebrew diagram showing
movements of the planets, days in the
months, and signs of the zodiac.**

Facing page 64
**A stopwatch by Breitling for use in
Radio and TV production. The outer
scale shows feet of 35 mm or 16 mm
film consumed.**

The author thanks horologist Brian Jackson of
Cambridge for the loan of reference works
from his library.

Contents

Introduction

Concepts of time

It is hard to imagine a world in which there is no measurement of time. And yet, compared with the history of mankind, time measurement as we know it is comparatively recent. However, observing and marking the passage of time in some way or other goes back to the dawn of the human race.

How might early humans have viewed their world? It is entirely believable that they saw themselves as a part of their environment, rather than somehow superior to it as we have come to think of ourselves. It is not altogether odd that the apparent harm we are doing to our environment in so many ways has at last turned us back to seeing ourselves as part of, and custodians of, our planet – a view which has long been held by American Indians, to mention but one culture.

So in the days of which we are thinking, humans were – as they still are – at one with nature. We are surrounded by cycles of birth, growth and death. People and animals live and die. So do plants, some on a shorter time scale, some on a longer. Who can fail to marvel at the great oak growing from the acorn? Without the benefit of our lofty understanding, early people would have seen the sun born and dying each day; the Moon doing the same in a different way over a longer period; the earth herself repeating the seasons over and over and over again.

All these cycles – animal, plant, solar, lunar and seasonal – have different characteristics, but the common theme is birth, growth and death. Exercising their reasoning power, early people must have seen themselves as having an increasing control over themselves and their environment, a control which they might – wittingly or otherwise – think could be extended from the real (making fire, growing crops, harnessing animals) to the pretend (making rain, calling up eclipses, conciliating the spirits).

By the time you reach that point, you will surely have some system of observing, measuring and recording, however simple it may seem by our standards. But it's not so simple, because it takes more than one lifetime of observation, measurement and recording to plan, build – and check – a structure such as Stonehenge in southern England.

Even seasonal variations take many cycles to work out if no one has done it before, and one is not yet certain what is coming next, and what one ought to be measuring, and how to measure and record it.

If you can somehow clear your mind of much of what you know and accept, think about how you would go about such a task. That night follows day is obvious. So you devise a way of marking off days – but don't forget that you haven't any knowledge of counting systems or notation; nor are you sure *why* you want to count days. Against your record of days you will start to note events (and you'll need a way of doing that) – but you don't yet know what events are significant, and what the significance is. Rain? Winds? Thunder and lightning? Earthquakes? Snow and ice? Eclipses? Buds bursting? The first cuckoo? Ripening fruit? Autumn mists? Eventually, it will become apparent that there are natural cycles of events – rivers flooding, rainy seasons, seasons of cold and heat, birds and animals giving birth to young, land dormant and land burgeoning, and so on. But it will take you some time to make sense of it if you start from scratch.

Some commentators have rejected the day as the first-recorded time interval, on the grounds that the tallies involved would become very large. But that is a view of hindsight, and straightforward counting (with notches, twigs or pebbles for example) is simple compared with all the other advances of observation and recording needed to extract some order from nature.

Now that we are counting days, and have presumably established that there is an annual cycle, we can

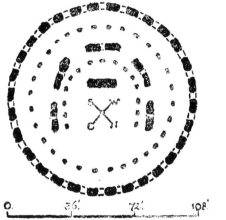

An idealized picture of Stonehenge. The stones are aligned with the rising of the sun on Midsummer Day, assumed to happen at the top of the drawing. Stonehenge dates back to about 2600 BC.

apply the information not only to cyclical events, but to measuring gross lengths of time. Lifespans can be measured, and then more advanced concepts can be introduced – such as the number of years which should elapse before a child passes to adulthood.

It is hardly likely that someone making such observations would neglect the Moon and the stars; and we know that astronomy (and astrology) are of great antiquity. The art of stellar navigation at sea is not part of this book. Observing the constellations, identifying what we now call planets, and groping towards a mechanism to explain the apparent relative movements of everything is a parallel story. However, maritime affairs have had their place in driving the technology of timekeeping, as we shall see.

Travellers in antique lands

A prime reason for travelling is to exchange goods – either for other goods (barter) or, later, for money. Focusing trade on one place – a market – is obviously a good idea, and arriving on market day is important.

At first, travel distances must have been measured in terms of time – as, indeed, they often are today, for time takes into account different degrees of difficulty on the journey. What supplies do I need for a longer journey? If I'm going there and back, when should I expect – and be expected – to return again? If I haven't arrived in the expected number of days, where have I gone wrong?

All these questions are a consequence of measuring the passage of time; once you have come to grips with recognizing and measuring a particular period of time, that ability will give you ideas for the next subdivision.

There are two distinct ways of setting up a calendar for regulating such activities: by counting the days, or by noting the phases of the Moon. That these two ways came together is shown by our calendar, which combines the week – arbitrarily of seven days – with the month (literally 'moonth'). The Moon is not a wholly satisfactory timekeeper, for its cycle bears no exact relationship to a number of days (and there's no reason why it should).

The calendar

Natural cycles

Because time is related to the motions of heavenly bodies (and *vice versa*), it is natural that astronomy and timekeeping should have developed hand in hand.

The problem is that the observable periods of the Sun, the Moon and the year are not related to one another. The alternation of night and day is clear enough, but the periods of light and darkness are not generally of equal length from one day to the next. The phases of the Moon are equally obvious, but the relationships of the lunar cycle to the day, and between one lunar cycle and the next, are disconnected. Yet the moon, giving – as it does when full – some nocturnal illumination, must have been of particular importance to those without means of portable artificial light – the expressions 'Hunter's Moon' and 'Harvest Moon' speak for themselves.

The year is a less obvious period of time than either the day or the lunar cycle, manifested by the procession of seasons, the behaviour of plants and animals, and large events such as tides and floods.

Whatever the difficulties, some means of recording the passage of time is needed so that people can meet nature by being ready for natural events such as rains, floods, the growing season, and the arrival of herds of animals, shoals of fish or flocks of birds. For example, the palolo worm (*Eunice viridis*) swarms to the surface of the South Pacific during the third quarter of the November Moon. How it tells the time is not clear, but the inhabitants of the surrounding islands lie in wait for it and gather a rich harvest every year.

We do not know whether religion arose as a result of a need for an explanation of the environment in its widest sense, as a reconciliation for death, as a means of securing power by those who took the lead, or as a combination of all these and other factors. One manifestation of religion is certainly bound up with natural events – as witnessed by the many festivals celebrated

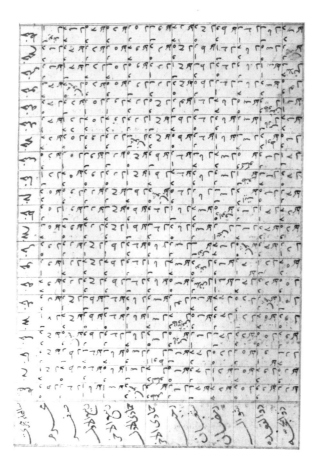

A Muslim lunar calendar based on the date of the 'Hijra' – the migration of the Prophet Muhammad from Mecca to Medina. The Muslim era started in the year of the Hijra (in Latin *Anno Hegiræ*, abbreviated to AH); 1 Muharram AH (the first day of the first month of the first year) corresponds to 16 July 622 AD. In 1992, 622 AD was 1370 years ago, but 2 July 1992 AD corresponds to 1 Muharram 1413 AH – a striking example of the discrepancy between the lunar cycle and the solar year.

Iran and Afghanistan use a solar calendar based on the Hijra, so their year is generally 622 less than the Gregorian year.

Left
3rd-century Mayan 'stele' – a stone post – erected to mark the passage of a five-year period.

by the many religions and systems of belief throughout the world – either to make something happen, or to commemorate or give thanks for something that has happened within legendary, if not living, memory.

The very existence of a calendar implies a cyclical view; of similar times (flood, harvest *etc*) returning; of similar festivals to be celebrated. Certainly 'knowledge is power' and the need for early calendars was driven by such activities as agriculture, religion, and public life. Each group of calendar users had its own needs and emphases, and many different systems arose in those early days – indeed, having a calendar helped to establish a group's identity and independence.

Early calendars

Early calendars had to do their best to reconcile the day, the lunar month and the year. The problem is that the lunar year is about 354.3 days, whereas the time taken for the earth to complete an orbit of the sun is about 11 days longer than that. In the lunar Muslim calendar the effect is that the date on which the Muslim New Year falls moves steadily backwards through the solar calendar. The Hebrew calendar inserts an extra month in the 3rd, 6th, 11th, 14th, 17th and 19th years of the metonic cycle – named after the Greek astronomer Meton (5th century BC) who saw, by continuous observation over many years, that the new Moon falls on the same day of the (7-day) week every 19 years.

Because of the lack of relationship between observed periods of time, the history of our calendar is one of confusion and compromise. Today, unless we are astronomers, we are almost certainly unaware of the difference between the astronomical and the civil (calendar) year. The former is (about) 365 days 5 hours, 8 minutes 46 seconds; the latter is arranged so that its Midsummer Day (now) falls on 24 June.

The Ancient Egyptians developed a 12-month year with three seasons (flood, seed-time and harvest), each consisting of four lunar months. From time to time, an extra lunar month was inserted to reconcile the phases of the moon with calendar reality. Observation established that a year was about 365 days long, and as long ago as 3000 BC the Egyptians introduced a second calendar – running in parallel with the lunar calendar – still with three seasons, but now with 12 months of

Like the Muslim calendar, the Hebrew calendar is lunar. It takes its date of origin as an important religious event – in this case, the Creation, on 7 October 3761BCE (Before the Common Era). However, extra months have kept the Hebrew and solar calendars more or less in step: 3761 + 1992 = 5753, and 9 September 1992 is 1 Tishri 5752, the first day of the Hebrew year.

This diagram, from a Hebrew manuscript, shows the movements of the planets at different times of the year, the number of days to each month, and the names of the signs of the zodiac.

30 days each, and five supplementary days before the start of the new year.

The year was still about ¼ day short, so a third lunar calendar was later introduced to remedy the discrepancy. The ancient Egyptians thus had three calendars to deal with at the same time.

The Babylonians and ancient Greeks also devised lunar calendars of increasing degrees of complexity as they developed their understanding of the mechanisms of the heavens. The Greeks introduced and modified calendars which were independent of the lunar cycle until they adopted the Roman 'Julian' calendar (named after Julius Cæsar) with Christianity in the 4th century AD.

The Roman calendar

The Roman calendar forms the basis of the calendar we use today. Legend has it that it was laid down by Romulus, founder of Rome in 753 BC – another example of an important event marking the start of reckoning. Roman years were reckoned AUC (*ab urbe condita* – that is, from the founding of the City [of Rome]). Romulus provided six months of 30 days (April, June, Sextilis, September, November and December) and four of 31 days (March, May, Quintilis and October). However, a year of 304 days is far too short, so in the reign of Numa Pompilius (715? – 673? BC) February was added to the end of the year, and January to the beginning. In 452 BC there was a rearrangement; the month of February was placed after January, and the year now had six months of 29 days and six of 30. However, this made 354 days – which for some reason was considered to be unlucky – so another day was added to make it 355.

In due course, it was observed that this year was still too short, so the month of Mercedinus, consisting alternately of 22 and 23 days, was inserted between 23 and 24 February in alternate years. This made the four-year cycle total 355 + 378 + 355 + 379 days = 1465 days, giving an average year of 366.25 days – about a day too long. It was therefore decreed that, in every third period of eight years, three, instead of four, months – each of 22 days – should be inserted, making the average year 365.25 days. As we know, there would have been simpler ways of achieving this result, but creeping tradition is a hard master.

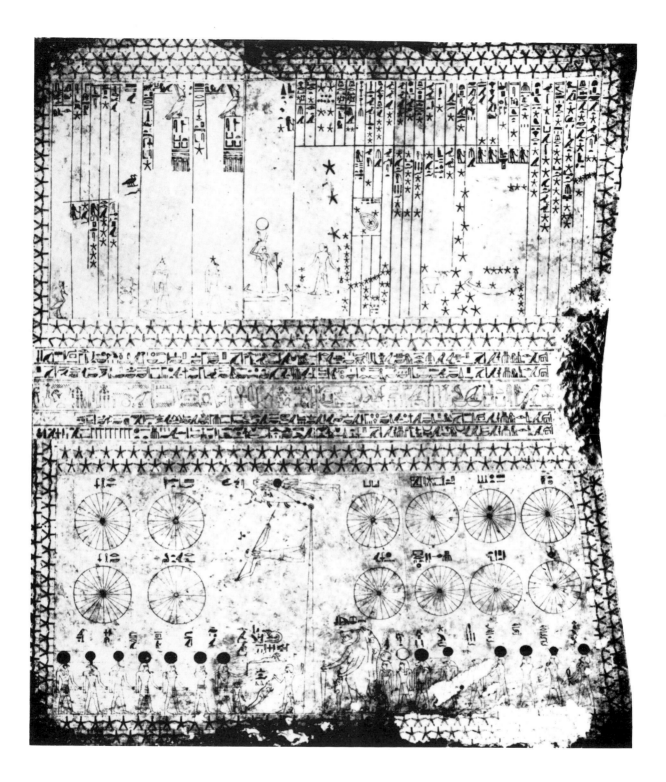

The astronomical ceiling in the tomb of Senmut, showing one of the numerous ancient Egyptian arrangements for relating the appearance of the heavens to the passage of time.

However, the calendar was not as well regulated as we have described, and those in charge manipulated it for their own ends. The result was that, by the time Julius Cæsar took charge, the 'equinoxes' (days of 12 hours light and 12 hours darkness) were three months out of step with the calendar.

Cæsar built on the calendar he inherited, abolishing Mercedinus and decreeing that the alternate months January, March, May, Quintilis, September, and November should have 31 days, and the rest 30 – except for February, which would usually have 29 days, but 30 days every fourth year (sequence A opposite). A four-year cycle now consisted of 1461 days, giving an average year of 365.25 days.

The addition of January and February at the beginning of the year made the 5th to the 10th months (Quintilis, Sextilis, September, October, November and December) the 7th to the 12th months. In spite of the fact that their names were derived from the Latin 5, 6, 7, 8, 9, 10, the shift does not appear to have given any concern.

In 44 BC, Quintilis was renamed July in honour of Julius Cæsar, and in 8 BC Sextilis became August to honour Augustus Cæsar. At the same time, a day was taken from February and added to the 30-day Sextilis so that Augustus's month should be as long as that of Julius (sequence B). It was considered that there should not be three 31-day months together, so September and November were given 30 days, and February had 28, or 29 in a leap year (sequence C).

The development of the Julian calendar

In that form (sequence C), the system of the Julian calendar was – and is – almost good enough to be used today. But it is still 11 minutes 14 seconds too long which, though it seems a small discrepancy, amounts to 18.72 hours over 100 years. The discrepancy was noticed by those responsible for fixing the date of Easter, following a method of calculation settled by the Ecumenical Council held at Nicea in Asia Minor in 325 AD. The Council also decided the year of the birth of Jesus, and – with a stroke of practicality – determined that leap years would be those divisible by four without remainder.

Saints' Days in March – a calendar page in a fifteenth-century manuscript.

The Gregorian calendar

The 19th Ecumenical Council of the Church of Rome – the Council of Trent (1545–63) – found that the spring equinox (fixed by the Council of Nicea as 21 March) had fallen back to 11 March, and authorized the Pope to look into the matter. It was Pope Gregory XIII who, in 1582, decreed that the day after 4 October 1582 would be 15 October 1582. He also decreed that, if a century year was not divisible by 400, it should not be a leap year. This omission of (in effect) three days every 400 years gives an error of about two-and-a-half days in 10,000 years – which could be corrected if years divisible by 4,000 (at present leap years) were not leap years.

Today, we work to the Gregorian calendar, though it was not adopted until 1752, when the Calendar New Style Act (1750) came into force, and Britain and the

Sequences A, B and C of the developing Roman calendar.

A		B		C	
Month	Days	Month	Days	Month	Days
January	31	January	31	January	31
February	29(30)	February	28(29)	February	28(29)
March	31	March	31	March	31
April	30	April	30	April	30
May	31	May	31	May	31
June	30	June	30	June	30
Quintilis	31	July	31	July	1
Sextilis	30	August	31	August	31
September	31	September	31	September	30
October	30	October	30	October	31
November	31	November	31	November	30
December	30	December	30	December	31

The French Republican calendar was devised in 1792 by Revolutionaries dedicated to renewing everything. It petered out in 1805.

Month	Meaning	Dates
Vendémaire	Vintage	22 September – 21 October
Brumaire	Mist	22 October – 20 November
Frimaire	Frost	21 November – 20 December
Nivôse	Snow	21 December – 19 January
Pluviôse	Rain	20 January – 18 February
Ventôse	Wind	19 February – 20 March
Germinal	Seedtime	21 March – 19 April
Floréal	Blossom	20 April – 19 May
Prairial	Meadow	20 May – 18 June
Messidor	Harvest	19 June – 18 July
Thermidor	Heat	19 July – 18 August
Fructidor	Fruit	19 August – 16 September

Five festival days: Virtue, Genius, Labour, Opinion & Rewards

17 – 21 September

British Dominions omitted the days 3–13 September from their calendars – for another 24 hours' discrepancy had accrued since 1582. The cry from those who failed to understand what had happened was: 'Give us back our eleven days'. At the same time as the calendar was adjusted, New Year's Day in England and Wales, which had been 25 March, became 1 January, a change that had taken place in Scotland in 1600. North America, being a British Dominion at the time, changed its calendar in 1752; Alaskan Territory clung to the Julian Calendar until 1867, in which year it was transferred from Russia to the United States.

The week

The seven-day week is of great antiquity, and is best known via its Hebrew roots in the story of the Creation, though it appears in other cultures as well. Why seven days? Perhaps because of its near relationship to the 29.5 days the Moon takes to circle the Earth. In the Hebrew and Muslim calendars, the lunar month begins at sunset of the day when the new moon is first seen after sunset. The months in these calendars are sometimes 29, sometimes 30, days.

The length of the week, then, is not based on any celestial phenomenon; it seems to have arisen as the result of a natural need to impose a certain periodicity on life, and to take a day of rest from time to time. 'Weeks' of as few as five – or as many as 10 – days have been recorded in different calendars. The seven-day week coincides with the Jewish story of the Creation; if God needs to rest on the seventh day, how much more does man?

Seven, moreover, is a magic number – hence the seven 'stars' in the sky: Saturn, Sun, Moon, Mars, Mercury, Jupiter, Venus. We derive the names of Saturday, Sunday and Monday from Saturn, the Sun and the Moon; then, from the Norse language, Tuesday (Tiu), Wednesday (Woden), Thursday (Thor) and Friday (Freya). The Græco–Roman influence is clearer in French: Lundi (la Lune), Mardi (Mars), Mercredi (Mercury), Jeudi (Jupiter) and Vendredi (Venus).

Shadows, water and sand

Movements of the Sun

Devising a calendar to measure the passage of time by the day, week, month and year is a very different task from that of subdividing the day into hours. Certainly, the same stick as that used to determine the direction of the rising Sun may be used to cast a shadow which moves throughout the day, but how to divide the day is not obvious.

It is easier to mark the passage of the shadow and divide its total arc into a number of equal angles than it is to make the angle (= period) of the subdivision (hour) equal from day to day.

Another method of measuring the apparent position of the Sun (supposing, as always, that it's visible) is to use a shadow stick – or 'gnomon' – as in a sundial. However, instead of measuring the progress of the day, note the position of the shadow when the Sun is overhead from one day to the next. Alternatively, take sightings of the rising Sun, and note its apparent

Shadow boards are still used in Upper Egypt for measuring the time taken to perform tasks, or for timing the distribution of water for irrigation. A water clock is used to mark the movement of the shadow. *(Science Museum)*

Samuel Smiles tells the following story of George Stephenson, the railway pioneer, and his son Robert. While Robert was still at school, his father proposed to him during the holidays that he should construct a sun-dial, to be placed over their cottage door at West Moor. 'I expostulated with him at first', said Robert, 'that I had not learnt sufficient astronomy and mathematics to enable me to make the necessary calculations. But he would have no denial. "The thing is to be done," said he, "so just set about it at once." Well; we got a *Ferguson's Astronomy*, and studied the subject together. Many a sore head I had while making the necessary calculations to adapt the dial to the latitude of Killingworth. But at length it was fairly drawn out on paper, and then my father got a stone, and we hewed, and carved, and polished it, until we made a very respectable dial of it; and there it is ... still quietly numbering the hours when the Sun is shining.'

The Stephensons' sundial is there to this day; it is dated August 11th MDMCCCVI (1806).

movement throughout the year.

It is obvious that, within the day, the Sun appears to rise in one direction and sink in the other. It therefore describes an arc across the heavens, and this governs the direction of shadows cast by objects. A primitive sundial can thus be set up, to give some measure of when communal events – such as mealtimes – should take place, and to give an idea of how much longer it will be light enough to see what one is doing, when to return from the hunting expedition, and so on.

The sundial has reached some degree of sophistication over the years, and constructing one is an interesting exercise. Ferguson's *Astronomy* gives the following instructions. Make four or five concentric circles, a quarter of an inch from one another, on a flat stone. Fix a pin perpendicularly in the centre, of such a length that its whole shadow falls within the innermost circle for at least four hours in the middle of the day. Set the stone level, in a place where the Sun shines from, say, eight in the morning till four in the afternoon. Watch the times in the morning when the tip of the shortening shadow just touches the circles, and mark those places. Then, in the afternoon of the same day, watch the lengthening shadow and, where its tip just touches the circles, make marks also. With a pair of compasses, find exactly the middle point between the two marks on any one circle, and draw a straight line from the centre to that point. The line will be covered by the shadow of the pin at noon.

The Earth rotates on its axis, which (by definition) passes through its North and South Poles. A solar day is measured by the interval between two noons, and we divide it into 24 hours. On that time scale, however, the Earth takes 23 hours 56 minutes 4.1 seconds to complete one revolution, because as well as revolving on its own axis it is moving around the Sun, and therefore it 'gets there sooner'.

The latitude at which the Sun is directly overhead at noon changes because of the above factors. From the shortest day (21 December) to the longest (21 June), the Sun appears to travel a little further north each day; for the other half of the year, it appears to move south. The vernal equinox – the day in spring when night and day are of equal length – is on 21 March; the autumnal equinox is 23 September. To be 'accurate', a sundial has to be designed for its latitude of use; moreover, sundials take no notice of the step changes in human time

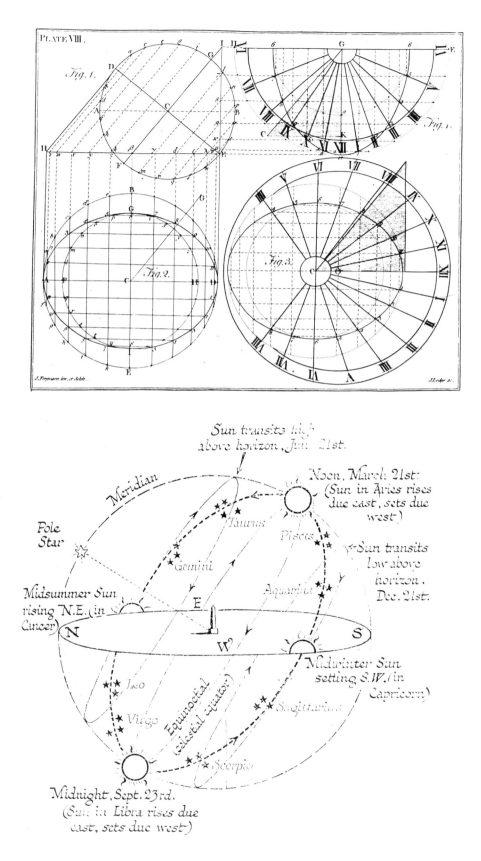

A plate from Ferguson's *Select Mechanical Exercise.*

'The plane of the ecliptic' is the name given to that plane in space in which the Earth's orbit lies. The axis of the Earth is not at a right angle to the plane of the ecliptic; it is tilted at an angle of 23°26' with respect to it; moreover, the orbit is not circular. None of this behaviour pays any respect to an emerging civilization struggling to make sense of the heavens!

L Hogben, *Science for the Citizen*

which now by convention changes by one hour for every 15 of longitude. And sundials bought for garden decoration may be so inaccurate that they bring the whole art of 'dialling' into disrepute.

The water clock or clepsydra

The progress of a shadow on a sundial measures time variably; the water clock or 'clepsydra' measures it as a constant flow. The ancient Egyptians were preoccupied with death and resurrection, and the behaviour of the Sun (dying each evening) provided constant work for their priests who wished to ensure its continuing morning reappearance. During the day, the Sun itself could cast a shadow to mark the passage of time; in the hours of darkness the clepsydra performed that function.

The principle of the clepsydra is simple: water flows (or drips) from a hole in a reservoir. The level of the water drops, and the passage of time can be read from marks inside the reservoir. Alternatively, the water flows from the reservoir into another vessel, and the passage of time is read from the rising level of the surface. Another method is to place a perforated bowl in the reservoir, and observe the bowl sinking as the water leaks in; alternatively, a calibrated air-filled vessel may sink as the air leaks out.

Cast of Karnak clepsydra of Amenhotep III (1415–1380BC). The year was divided into 12 months, and the days – as opposed to the nights – divided into twelve hours. The hours thus varied in length; inside the bowl there are twelve sets of dots to mark the passing of the hours – one set for each month.

Science Museum

Two water clocks described in 1634.
Left:
A vessel marked in intervals of time gradually sinks into water as the air trapped beneath it leaks through a hole in its top.
Right:
A floating model of a skeleton points to a time scale as water flows in to its bath.

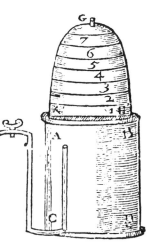

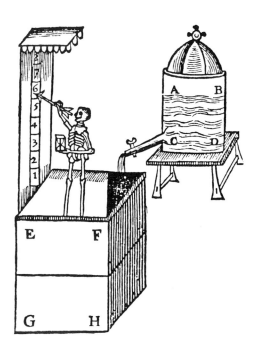

15

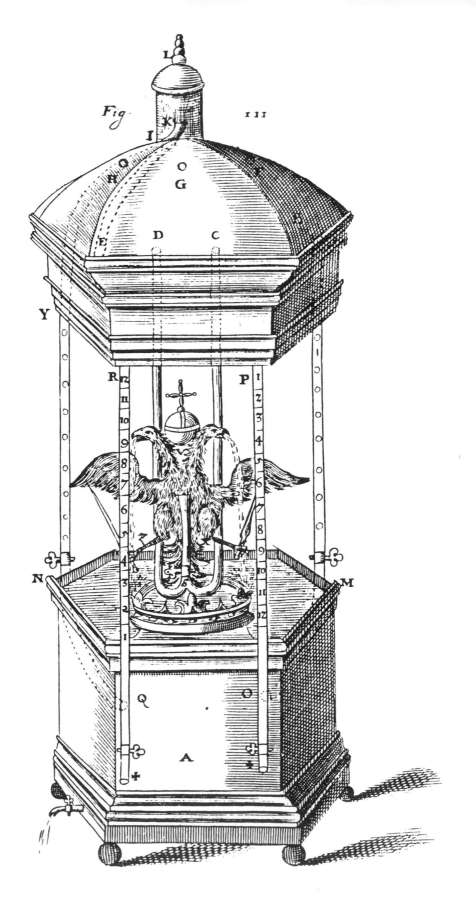

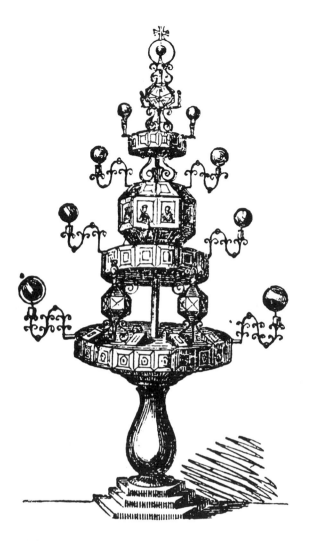

'Dials' at Whitehall, 1669. How King Charles II's timekeeper worked is not clear. Britten writes: 'This curious erection had no covering; exposure to the elements and other destroying influences led to its speedy decay and subsequent demolition.' However, it is elsewhere reported that this construction of glass globes became the focus of the well-known rake John Wilmot, 2nd Earl of Rochester, and his companions as they returned from some drunken gathering. Stung by its impudence, they laid about it with their sticks and smashed it to smithereens. The King was not amused.

Left
A fanciful 17th-century water clock hydropneumatic device which was more use as a conversation piece than a timekeeper.

The clepsydra clearly offers scope for mechanical ingenuity; later versions show a float rising in a filling vessel, operating a rack-and-pinion mechanism which causes a rotating hand to indicate the hour. A similar effect is achieved with a cord and counterweight, and the float driving a figure which points to the hour on a vertical scale. If the clepsydra is to measure variable hours – where the day is divided into equal divisions according to the interval between sunrise and sunset – another water-driven mechanism rotates the vertical scale to which the figure points.

Clepsydras are relatively easy to make and maintain, and were used until the 18th century. The rate of flow of a constant head of water through a given hole is generally affected only by changing temperature, so the accuracy of their timekeeping (they would be set right by the Sun), once adjusted, would be consistent.

Much ingenuity was devoted to the clepsydra, and some models are said to have marked the hours with bells or trumpets. Procopius, the 6th-century historian, tells of a mechanism in the market place at Gaza – a clock in a small, temple-like building, which had 12 decorated doors corresponding to the hours of the day, and 12 plain doors for the hours of the night. A figure of the sun god moved across to mark the passage of time and, on each hour, a figure of Hercules struck a bell and the appropriate door opened and figures acted out one of the 12 Herculean labours.

However, before we get too carried away with the Lost Knowledge of the Ancients, let us pause to ponder on the possibility that there might have been someone inside watching a clepsydra or other meter and pulling the strings accordingly. And if the clock was the standard of time, nobody could say it was 'wrong'!

The sandglass

The sandglass – or hourglass – is of great antiquity. The ancient Greeks used it – as we do today – for measuring cooking times (especially of boiling eggs). They also used it for timing speeches and, in conjunction with the log (a piece of wood on the end of a string knotted at equal intervals), for measuring the speed of a boat.

The sandglass might be thought of as a clepsydra filled with a solid material; however, the sand must be kept perfectly dry and, since the sand must also be visible, it has to be sealed into a transparent container.

This illustration from a German–Hebrew Pentateuch of 1395 shows a sandglass hanging in a schoolroom. Sandglasses were used in this way until at least the 17th century – presumably either to tell the student (or the teacher) when it was time for a break, or to enable the teacher to charge fairly for his work.

It is said that a French monk, Luitprand of Chartres, 'resuscitated the art of blowing glass' at the end of the 8th century, enabling this very accurate timekeeper to enjoy a revival. When the practice of 'payment by the hour' was introduced in the 14th century, the sandglass was often used for timing work.

The particles in a sandglass – which may be powdered egg shell, marble-dust, or even sand – must be carefully prepared and sieved so that they are all about the same size. The hole through which they flow should be about ten times the particle diameter. The rate of flow is independent of the amount of material waiting to pass through, and the air displaced from the lower vessel finds its way through the particles to the upper vessel, so there is no back pressure.

The aesthetic properties of the sandglass have also been remarked: its silent running, and the shape of the growing conical pile below echoing that of the conical depression above.

The earliest known picture of a sandglass appears in a fresco of about 1340 by Ambrogio Lorenzetti in the Sala della Pace in Siena, Italy. The sandglass (or hourglass) was used in English churches to time sermons for a couple of hundred years or so, and there are many accounts of the preacher ascending the pulpit steps and deliberately turning the sandglass over.

Burning time

A burning wick of one sort or another affords another means of measuring the passage of time. The calibrated candle is well known, and can still to be bought, usually for decorative purposes. An interesting use is 'selling by the candle'. At an auction, a pin is pushed into the side of a burning candle, and bidding proceeds until the pin drops, whereupon the current bidder is deemed to have bought the lot. The technique is mentioned by the poet 17th–century poet John Milton, and is known to have been practised in England at the end of the 19th century.

Another way of measuring time by burning is to be found in the oil lamp timekeeper. As the wick burns, the level of the oil in the glass reservoir falls; calibrations on the glass indicate the passage of the hours of darkness.

The Chinese used burning incense in pierced boxes of various shapes, both plain and symbolic. Various methods were developed for indicating the passage of time. In one 18th or early 19th century model, a dragon boat contains a pewter tray in which an incense stick is burnt. Stretched across the incense stick are six threads, each held in place by a small spherical weight at each end. As the stick burns, the threads burn through and the balls fall on to a sounding plate beneath. The spacing of the threads determines the time intervals measured.

Although all the devices described measured a period of time accurately whenever they were set going, such periods were not generally related to a universally-accepted scale of time. It was better that the device should be fit for its particular purpose than that it should relate to an absolute measurement, or be comparable with someone else's. This may seem odd in the light of our modern understanding of time, but even now we care not how long the sand runs as long as the egg is done to our liking.

Marking the minutes

The escapement

The sandglass, clepsydra and sundial are all *analogue* devices, in that there is a steady movement of the sand, water or shadow. As soon as we apply our ingenuity to mechanical devices for marking the passage of time we encounter the *digital* device, which moves in discrete steps: 1 – 2 – 3 . . . This in turn leads to the idea of time as a succession of short periods – such as the second. The regulating device which marks these short periods is the *escapement*, and the means of regulating the escapement is some sort of device swinging first one way, then the other – such as a pendulum.

Note that, though the escapement is a digital mechanism, the hands of the clock sweep round an analogue display. We can observe our attitudes of mind towards time by comparing our mental processes when we want to catch an afternoon train leaving at 3:27, and we look at the hands on a dial indicating five minutes past three, or a digital display reading 1505.

The celestial river (T'ien ho)

The Great Steelyard Clepsydra of Kêng Hsün and Yüwên Khai dates from 610 AD, and is thought to have worked as follows. At one end of a pivoted arm (or 'steelyard') is a container floating in a reservoir, from which it is filled by a syphon (the weight of the container keeps the level of water in it lower than that in the surrounding reservoir). When the reservoir is nearly full, an attendant moves a counterweight on the arm so that the container is lifted and the water flows back to the reservoir through the syphon. The whole is contained in a box to protect it from disturbance.

The mechanism has been described as an escapement, though it does need an operator. However, it marks a tradition of Chinese water escapements leading to that built by I Hsing and Liang Ling-tsan in 725 AD, and another fully described by the Imperial

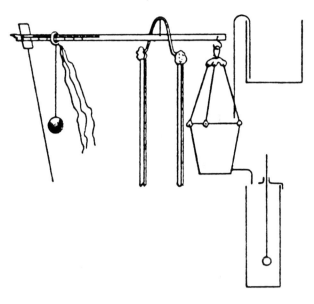

There are many ways of arranging a steelyard clepsydra. This diagram shows a horizontal steelyard arm with its counterweight on the left and bucket on the right. Water flows through a syphon from the top vessel, and empties through a tap into the bottom vessel.

J Needham *Science and Civilisation in China*

19

Tutor Su-Sung in his book *Hsin I Hsiang Fa Yao* in 1088 AD. Su-Sung's Astronomical and Time-telling Tower was 30 feet high with a water-powered driving wheel and an escapement to control the timekeeping.

Su-Sung's water wheel has 36 buckets, which fill one after the other. The filling bucket is restrained by an escape mechanism; it tilts when full and the wheel turns ten degrees of arc. It takes 24 seconds for a bucket to fill; one revolution of the main shaft therefore takes 14 minutes 24 seconds. This seems less odd when we realize that it is equivalent to 100 revolutions in a day (24 hours). The main shaft of the escape wheel drives the various time-telling elements in the upper part of the tower.

Early clockwork

The origins of the first mechanical clocks in Europe are not known, though it likely that they were devised to mark the Divine Offices in monasteries. This may have led to some confusion, since the word 'clock' is derived from the root we see in the French 'cloche', a bell, re-flecting the early 'clocks' in which a bell marked the passage of time. Equally, when the Latin word 'horolo-gium' is used in a mediæval text, it might mean a clep-sydra, sundial, mechanical clock or bell. However, the bell of the reported 'horologium' or 'clock' may have been struck by a monk watching a clepsydra, hour-glass, candle or other measuring device – remember the Gaza clock – and was not necessarily a wholly-automatic device. It has been suggested that a dearth of reliable (or willing) human timekeepers may have spurred the development of the mechanical clock.

The birth of the mechanical escapement clock is also unrecorded, though it is likely that many craftsmen were experimenting with such devices in various centres in the 13th century. A mechanical clock is said to have been built by the monks of Dunstable Priory, Bedfordshire, in 1283. This predates other recorded clocks by several decades. An astronomical clock with automata and a dial set up in Norwich Cathedral is recorded in the Sacrist's Rolls for 1325.

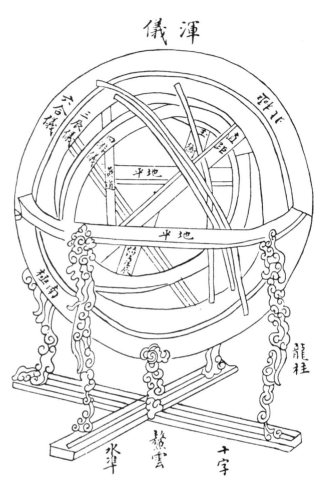

An 18th-century drawing of the armillary sphere which surmounted Su Sung's Astronomical and Time-telling Tower. The sphere is a skeleton celestial globe of circular hoops representing the equator, the ecliptic, and other imaginary features.

A modern reconstruction of Su Sung's water escapement, showing the 36 buckets and the filling and escape mechanism on the right.

Science Museum

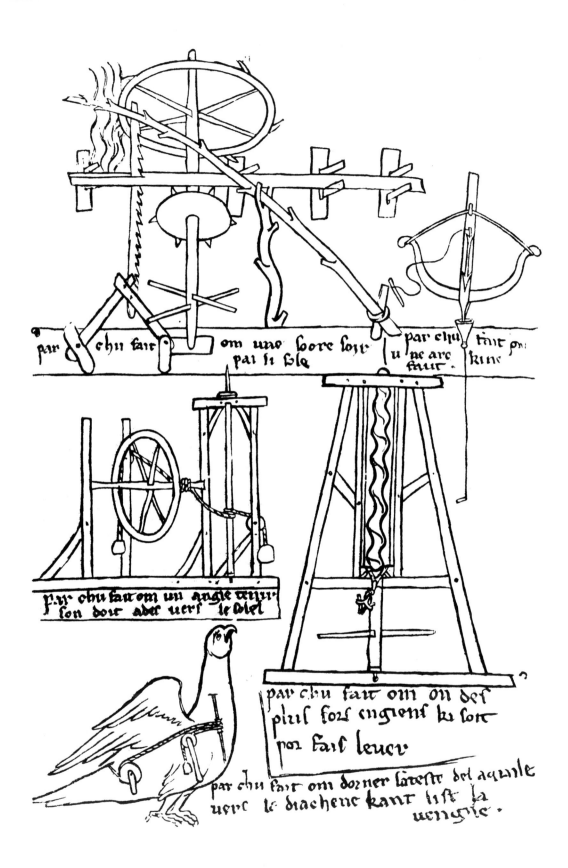

par chu fait om une hore soit
pai li isle

par chu fait on
u ne arc
fait.

kine

par chu fait om un angle tenir
son dort ader vers le solel

par chu fait om on des
plus fors engiens ki soit
por fait lever

par chu fait om dorner la teste del aquile
vers le diachene kant lisf la
uengile.

The verge escapement. The shaft of the toothed crown wheel – or escape wheel – is pulled round anticlockwise by a weight (not shown). However, the escape wheel cannot rotate freely because of the *pallets* on the vertical shaft. As shown, the top pallet is engaged by a tooth of the wheel. However, the pull of the weight will push the pallet to the left, turning the vertical shaft or *verge*. But then the lower pallet will engage with a bottom tooth, stopping the escape wheel and reversing the rotation of the verge. The rate at which the mechanism runs is determined by the driving weight, the length of the arms of the swinging *foliot*, and the size and positions of the weights thereon.

Left
This page from a Villard de Honnecourt sketchbook shows his rope escapement (centre, left) and an ornithological automaton. At the top is a machine for sawing a tree into planks.

Villard de Honnecourt

The sketchbook of the 13th–century French architect Villard (or Wilard) de Honnecourt (?1250–?1300) shows a sketch captioned: 'How to make an Angel point with his finger towards the Sun'. The Angel was apparently mounted on the roof of a church, and turned by a tensioned rope wound round his axle. In order to regulate the progress of the rope, it passes round another axle, and between two of the spokes of a wheel fastened to that axle. The weight pulls the wheel axle round until the rope is stopped by the one spoke of the wheel; it bounces and recoils and the rope slips until it is arrested by the other spoke, and so on. In principle, the approach is similar to the *verge* escapement described below. As drawn by de Honnecourt, it is difficult to see how the mechanism would have continued to oscillate in preference either to the mechanism grinding to a halt, or to the rope running through the wheel, causing the Angel on the roof to whirl round uncontrollably.

The verge escapement

By the middle of the 14th century, perhaps earlier, there were several weight-driven striking clocks in Europe (at Milan and Rouen, for example). Their escapement was a *crown wheel*, controlled by *pallets* on a vertical shaft, or *verge*. The verge oscillates about its vertical axis, and the pallets alternately engage with and release the teeth on the escapement, allowing it to turn half a tooth at a time. The oscillation of the verge is governed by a *foliot*: an arm with adjustable weights rotating this way and that.

If the arbor (axle) of the escape wheel is vertical, the verge may be governed by a (short) pendulum, though the resemblance to the later pendulum whose length regulates the clock with some accuracy is purely coincidental since the mechanism can be made to go faster by increasing the driving force, or decreasing the weight of the pendulum 'bob'.

De'Dondi

It must be remembered that, at a time when there was no tradition of clockmaking, the idea – let alone the execution – of a train of wheels to slow down the descent of a driving weight must have presented quite a problem. However, it is clear that manufacturing intri-

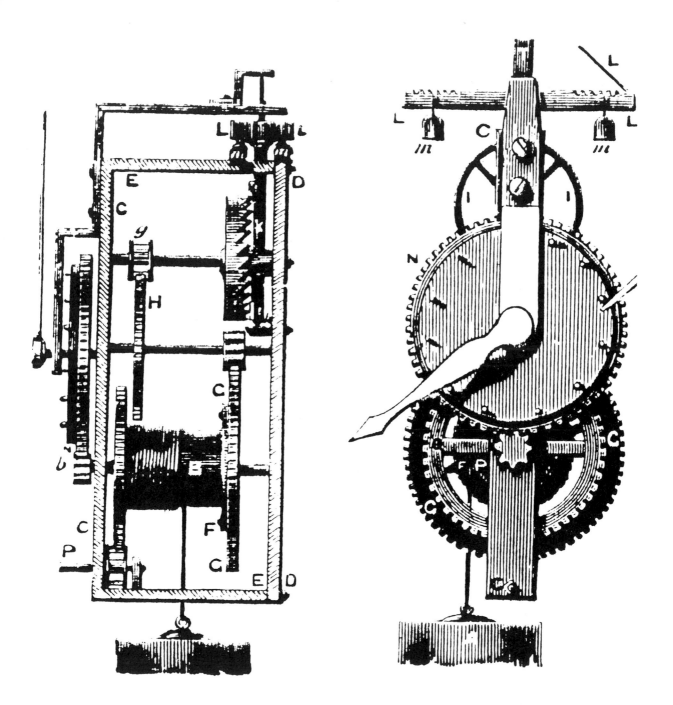

Diagrams of the mechanism of a late 14th-century striking clock by Henry Wieck, or de Vick. The escape wheel is clearly visible in the left-hand view and the foliot with its weights is seen in the right-hand view (see also page 32).

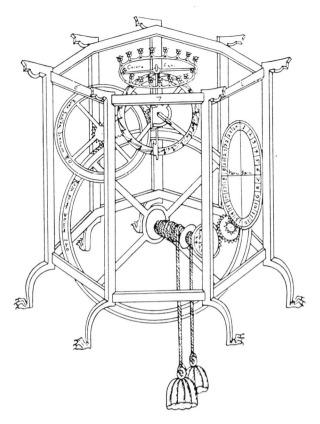

A general view of the main frame of De'Dondi's astronomical clock.

De'Dondi's astronomical clock: the gearing for the planet Mercury.

cate wheelwork was possible from an interesting early design (1348–64) of an astronomical clock by Giovanni De'Dondi, professor of astronomy in the University of Padua. De'Dondi's clock had dials showing 24-hour time, and lunar and planetary motions. It took him 16 years to build and seems to have disappeared in the 16th century – possibly destroyed by fire.

Fortunately De'Dondi left a detailed description from which replicas have been built. His clock shows the time on a revolving dial with fixed pointer (as was usual in those days); above the clock is an heptagonal planetarium relating to the heavenly positions of the Sun, Moon, Mercury, Venus, Mars, Jupiter and Saturn. Another dial shows the positions of the 'nodes' where the paths of Sun and Moon intersect, enabling eclipses to be predicted. An engraved drum shows the calendar day, date, number of hours of daylight, and the feast or saint to whom the day belongs.

Early turret clocks

In England, Richard of Wallingford, Abbot of St Albans, built a clock which was – at least partially – astronomical, between 1327 and his death in 1336. The earliest surviving mechanical clocks are from the cathedrals at Salisbury and Wells. According to the cathedral accounts, the Salisbury Cathedral clock was built in 1386. Towards the end of the 17th century it was converted – as many clocks were – to the newly-invented pendulum and anchor escapement system which we will describe later. The clock was discarded in 1884 and recommissioned – with a replica of its original verge and foliot escapement – in 1928; it now stands in the nave of the cathedral for all to see and admire.

The Wells Cathedral clock is thought to date from 1392 or earlier. Its escapement was converted in 1670; it told the time at Wells until 1835 and is now ticking away happily – and chiming the quarters and striking the hours – in the Science Museum, London. It is well worth a visit to see the state and quality of craftsmanship of the period; the frame at least is original, and its renewed parts are replicas of the old.

When at Wells, the clock boasted an astronomical dial, and on the hour figures of knights jousted, and a figure traditionally known as Jack Blandifer struck the hour on a bell; it is from this figure that striking 'Jacks'

Left

A 15th-century miniature of Richard of Wallingford, Abbot of St Albans, who devised an astronomical clock between 1327 and his death in 1336. The clock struck the hours, and the going train moved an astrolabe dial, as well as devices representing the motions of the sun, moon and perhaps the planets.

Right

The dial of the Wells clock, still to be seen in the cathedral. The works (*right below*) have long been in the care of the Science Museum, London.

Left

The Salisbury Cathedral clock, 1386. The uppermost toothed wheel is the anchor escapement (fitted in the 17th century). The mechanism was for striking only, and never had a dial.

take their name. Although the works of the original clock are in the Science Museum, the puppet display may still be seen in the cathedral, driven by a more modern mechanism.

It seems that both the Salisbury and the Wells clocks were commissioned by Erghum of Bruges, who was Bishop of Salisbury from 1375 to 1388, and of Bath and Wells from 1388 to 1400. In 1368, King Edward III invited three 'orlogiers' from Delft in the Netherlands – Johannes & Willelmus Vrieman and Johannes Lietuyt – to England, possibly to build a movement for his new clock tower at Westminster. One or all may have stayed on, and 'Peter Lightfoot', the horological wizard reputed to have built the clocks at Salisbury and Wells, might really have been Johannes Lietuyt.

These 'great clocks' – or turret clocks as we now call them – were not tremendously accurate by present-day standards. They had no dial or hands; their task was to proclaim the hours to the public by striking a bell. With the vagaries of the verge escapement, and the lack of a tradition of mechanical engineering at the time, it isn't surprising that such clocks might gain or lose 15 minutes in 24 hours – after all, that's an accuracy of about one per cent, and there was always the sundial by which to set them right.

As turret clocks were converted to pendulum regulation, and they could compete with sundials for accuracy, public dials became more common. The first public dial was displayed on the newly-built tower of Magdalen College, Oxford, in 1505.

Domestic clocks

There were a few domestic clocks in the 14th century; those that there were were kept mainly for prestige. If they had a dial, they would have but a single hand; the minute hand was a 17th-century addition. Early domestic clocks were weight driven – scaled-down turret clocks. There was little refinement in the wheelwork, so that very often they ran for less than 24 hours – a feature which anyone would surely find infuriating, and which may have made them more a status symbol than a useful timekeeper.

Domestic clocks with open sides – 'lantern', 'birdcage' or 'bedpost' clocks – appeared about 1600. They could be suspended from a hook, or stood on a wall bracket, and were wound by pulling on the opposite end of the rope or chain from that on which the weight hung – in the manner of the well-known tourist cuckoo clock. Early models used a foliot, sometimes an oscillating balance wheel, to control the escapement; later, the short bob pendulum came into use.

Audible time

Striking clocks

Sounding a preset alarm, or striking the hours, has long been one of the objects of a time-measuring mechanism – indeed, when clocks had no external dial, the sound of the bell was the only public announcement. Even when public dials appeared, the tradition of striking was maintained – perhaps to reassure the insomniac.

Early turret clocks marked the hours with a single

Black Forest clock of about 1740 with movement mainly of wood. The regulating 'pendulum' swings in front of the dial, an arrangement which cannot have many advantages, other than to make it easier to adjust the position of the bob.

A true figure of the famous *Clock of Strasbourg*

'A true figure of the famous Clock of Strasbourg' made in 1611. The giant clock of Strasbourg Cathedral, in France, was completed in 1354. It comprised a calendar of movable feasts, an astrolabe showing the relative positions of the Sun and the Moon, and an indication of the hours. Above the dial was a statue of the Virgin Mary, before whom the three Magi bowed at noon, at the same time as a cockerel (the only surviving part of this masterpiece) crowed and flapped its wings.

stroke on the bell – reasonably easy to arrange. Striking a sequence 1 – 2 – 3 . . . 10 – 11 – 12 seems to have come in by the mid-14th century.

Effective striking of the hours needs an independently-driven 'striking train' – a 'train' here being intermeshed gear wheels – to control a hammer which strikes a bell. The striking train is independent of the 'going' train, but it must be set off by it.

The earliest automatic striking mechanism is of a type controlled by a 'count wheel' or 'locking plate'. The circumference of this wheel carries twelve notches, in one of which a *detent* normally lies. The count

Left

The turret clock from Cassiobury in Hertfordshire was built about 1600. It is one of the very few to have avoided losing its verge escapement.

Early 16th-century domestic clock. It strikes the hours and quarters, and sounds an alarm.

An Italian alarm clock of the 14th or 15th century. The 'chapter ring' rotates once in 24 hours; a pin inserted in one of its holes sets off the alarm.

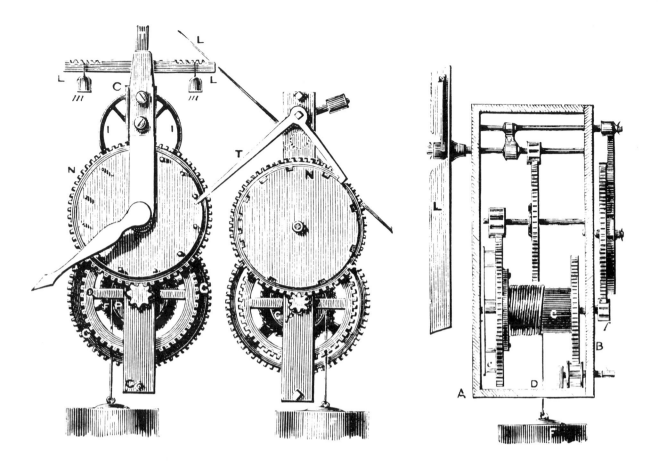

Another view of Henry de Vick's clock (see page 24). There are 12 pins on the wheel of the left-hand part of the mechanism. Every hour, one of the pins actuates the right-angled lever, releasing its other end from whichever notch of the count wheel it lies in. This sets the striking going until the next notch comes round and the lever drops into it. The part marked L on the side view is a *fly*, or fan, which regulates the speed of striking.

Striking mechanism of an eight-day clock. A is the rack with its 12 teeth; its tail B falls against the step of the snail C. D is the gathering pallet. The bell strikes once for each tooth of the rack gathered.

wheel is geared down from the wheel whose pins trip the hammer which strikes the bell. When the detent is released, the train is set in motion, and the bell strikes until the detent falls into the next notch on the count wheel. The notches are so spaced that the lengths of the 'lands' between them correspond to the times taken to strike one to twelve.

The disadvantage of this mechanism is that its counting is controlled by the striking train itself, so that the clock has to strike 1 – 2 – 3 . . . 10 – 11 – 12 in order. It cannot be made to repeat – that is, to strike the same hour more than once in succession according to the position of the hour hand. Moreover, if it gets out of step with the going train it can be tedious to reset it. What was needed was a mechanism which relates the number of strokes to the position of the hour hand.

The rack-and-snail mechanism invented by Revd Edward Barlow (né Booth; 1636–1716) in about 1676 meets that need. Each tooth of the rack is related to a stroke of the bell; when the striking train is in action, the teeth are 'gathered' one at a time, and each time the bell strikes. The number of rack teeth to be gathered is controlled by a stepped cam – the snail – fixed in relation to the hour hand. The tail of the rack can be made to fall against the snail as many times as you wish during most of the hour, always resulting in the same number of strokes. And if you move the hour hand round the dial, the snail turns with it, so that the count is always in step with the hour indicated.

Chiming

The term 'striking' is reserved for the hour strokes; chiming marks the quarters, or serves as a prelude to the strike. The sequence of more elaborate chiming is controlled by a pin barrel, sometimes driven by the striking train; sometimes with a train of its own. Some clocks have barrels which shift longitudinally, as in a musical box, so that different types of chimes can be called into play.

Intervals of time

One thing that the mechanical clock did was to measure time as a succession of equal intervals: tick – tock – tick – tock. It is difficult for us to imagine life without minutes and seconds; so much of modern life depends on time measurement that would have seemed unbelievably accurate in the days of those early mechanisms. Where would radio – and more particularly television – be without 'split-second' timing (one picture frame each twenty-fifth of a second)? Public transport is timed to the minute (whether or not this is a triumph of hope over experience). In the world of work, we observe the clock, we time our processes, we arrive at meetings 'on time' and we fidget about, conscious that 'time is money'. All this is a direct outcome of our ability to measure time; through the ages, time measurement has both reflected and led the need of the period.

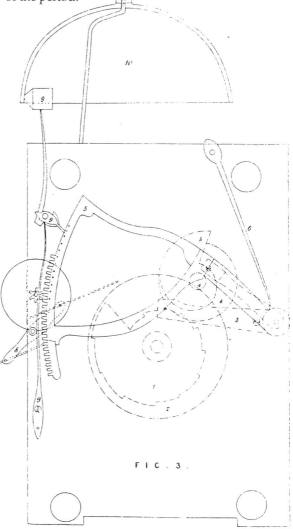

Eardley Norton (1760–94) was a maker of musical and astronomical clocks. His patent (No. 987 of 1771) is for striking work. The main snail is seen in the middle, and rotates with the hour hand (not shown).

The pendulum

Early ideas

Although the time of swing of the foliot escapement is constant, it is not easy to adjust, and the foliot therefore has a reputation as a poor timekeeper. The period is controlled by the position and mass of the weights on the arms of the foliot, by the stiffness of the suspension, and by the mass of the driving weight – as well as the friction in the system. Since all these had to be 'got right', there was little fine control – always remembering that the usual standard for setting the clock would be the more accurate sundial.

If time was to break away from dependence on natural phenomena and be brought into the workshop – or even the laboratory – a more controllable means of regulating the escapement was needed. This means was the pendulum. With hindsight, we can say that a free (unconstrained) pendulum, of a length of a metre or more, swinging through a small arc (a few degrees), is 'isochronous' – each of its swings occupies the same length of time. But it took a few hundred years for all those properties of the pendulum to be both understood and applied.

We should not be surprised to find that the first to think of using a pendulum to control a verge on a horizontal arbor was Leonardo da Vinci, though his idea does not appear to have been put into practice.

Although it didn't meet all the criteria of the pendulum proper, Leonardo's pendulum could have been a great deal more controllable and accurate than the verge escapement. In 1581, the Italian natural philosopher Galileo Galilei (1564–1642) discovered the isochronism of the swinging pendulum, reputedly by observing the period of a suspended lamp swinging

Drawing made by Leonardo da Vinci in about 1500, showing a pendulum-like element in the bottom left-hand corner. It has been suggested that, because of its layout, this mechanism is nothing to do with a clock, but Leonardo may have drawn it this way to show the parts and the gear ratios more clearly.

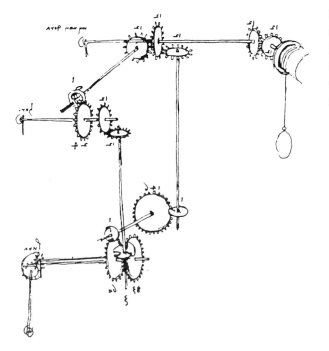

in the breeze in the Cathedral at Pisa. After much thought, he devised a mechanism, comprising pendulum and escape wheel, which would keep the pendulum swinging as it controlled the escapement. He (or his son Vincenzio) sketched a version of a suitable mechanism. Vincenzio was building a model of the mechanism when he died in 1649.

The pendulum harnessed

Galileo's idea of isochronism was developed by the astronomer Johann Hevel; Vincenzio's model was developed by Philip Treffler, clockmaker to Prince Leopold de Medici. However, both these attempts to turn the isochronous pendulum into a workable time-keeper were overshadowed by the work of the Dutch astronomer Christiaan Huygens (1629–95), who suggested a method of applying the pendulum to a clock mechanism in 1656. He assigned his invention to Salomon (or Samuel) Coster of the Hague, who patented it and built a spring-driven model the following year. In England, clocks to Huygens' pattern were built by the Fromanteel family of clockmakers, who sent John Fromanteel to work under Coster at The Hague and bring back the knowledge to London.

Huygens continued to carry out pendulum experiments, and found that in fact a wider swing took longer than a narrower one; for true isochronism the bob should swing in a 'cycloidal arc'. This he achieved by shaped 'cycloidal' cheeks between which the suspension threads of the pendulum swung. The effect of the cheeks is to reduce the length of the pendulum as its arc of swing increases. However, Huygens' models introduced other inaccuracies, so it was difficult to check the effect.

The pendulum applied to the verge escapement is not a particularly efficient combination, because the verge prevents the pendulum from swinging as freely as possible. The next improvement was the anchor escapement. Who invented this is not known – it might have been Robert Hooke, William Clement or Joseph Knibb. Clement built the earliest known clock to use the anchor escapement, which was set up at King's College, Cambridge, in 1671. Hooke, on the other hand, discovered about that time that the length of a one-second pendulum is 39.14 inches (994.16 mm). A pendulum of this length became known as the Royal

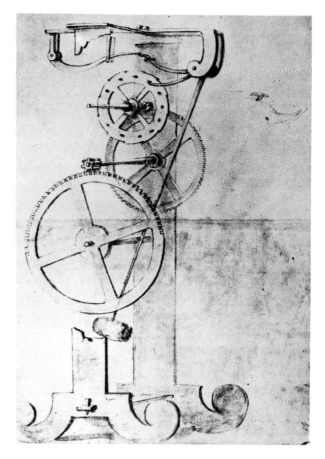

Vincenzio Galileo's drawing (1641) of his father's pendulum and escapement. The 'duck's beak' at the top is connected directly to the pendulum, and engages with pins on the escape wheel. At the same time, it releases the catch or 'pawl' from the escape wheel as shown.

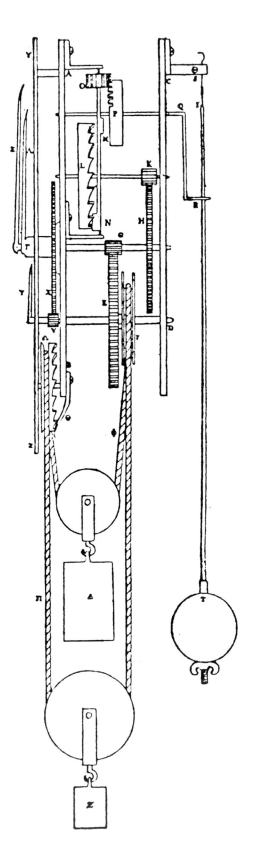

Diagrammatic side view of a clock mechanism fitted with Huygens' first pendulum, about 1658. The swing of the pendulum T is in the same plane as the escape wheel L. T drives L via 2:1 gearing. The gear F driving the verge shaft pinion O has no more teeth than it needs.

Right

Diagram of Huygens' clock with cycloidal cheeks T and escape wheel K in the horizontal plane. The form of the cheeks, and the two cords suspending the pendulum, are seen in perspective in Fig 2.

A diagram showing the effect of Huygens' cycloidal cheeks – ABC and CH. The pendulum is suspended from point C, but its suspension cord is controlled by the cheeks. In effect, the pendulum CBE becomes 'shorter' the further it swings from the vertical.

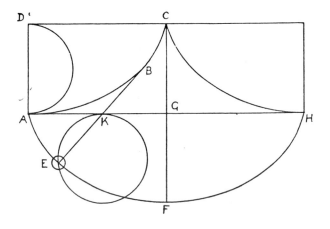

Pendulum, and paved the way for the minute, and then the second, hands to be introduced.

Once the mechanical clock, rather than the sundial, had become the prime timekeeper, clockmaking moved into a new era. From then on, accurate clocks with a heavy pendulum swinging through a small arc became the vogue. It was an obvious step for the mechanism to be enclosed, and for the clock to have its own case, rather than to stand on a wall bracket. This arrangement protected the pendulum from draughts and physical interference, and kept dust and dirt out of the mechanism; thus began the golden age of the long-case clock.

THE GOLDEN AGE OF THE LONG-CASE CLOCK
Developing the escapement

British clockmakers became supreme in the last quarter of the 17th century. The techniques of mechanical manufacture were established and developed; Robert Hooke, for example, invented a wheel-cutting machine about 1670. Effort could now be turned to increasing accuracy.

It will be noted that the second hand of a clock fitted with an anchor (or recoil) escapement moves back-

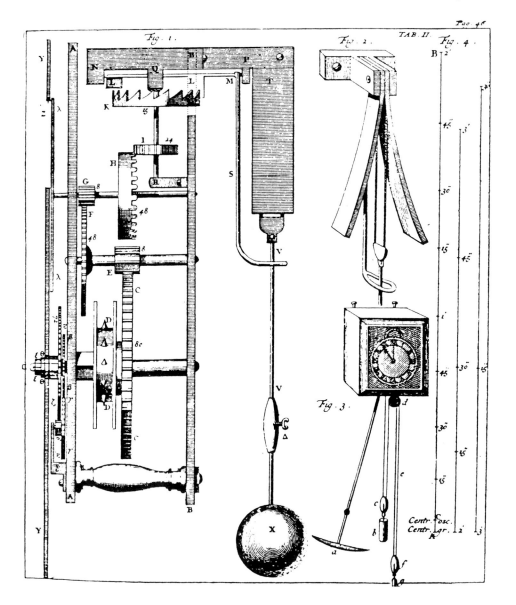

wards a little at each tick. In effect, the pendulum is pushing the movement backwards and raising the driving weight – perhaps by as much as 0.00001 inches! This deficiency was overcome by the 'dead beat' escapement, whose teeth are so shaped that there is no recoil as the pendulum (and hence the escapement) finishes its swing.

Who invented the dead beat escapement is unclear. It was long thought to be the work of George Graham in 1715, though there is evidence that the doyen of precision clockmakers, Thomas Tompion (1638–1713), had the idea well before the end of the 17th century, even if he didn't put it into practice until the beginning of the 18th.

An alteration in pendulum length of one thousandth of an inch is equivalent to one second per day; clockmakers therefore bent their minds to designing pendulums which would be less affected by changes in temperature. In 1722, George Graham produced a pendulum bob weighted with mercury; the idea was that, as the changing temperature caused the rod to change in length, the mass of mercury would be similarly affected, but in such a way as to keep the effective length of the pendulum (*ie* between the point of suspension and the centre of gravity of the bob) constant.

John Harrison produced his grid-iron pendulum in 1725: here, as the iron rods expanded downwards, the brass ones expanded upwards. At the end of the 19th century Dr Charles-Edouard Guillaume (1861–1938) invented Invar, an alloy of iron and nickel with a very low coefficient of expansion. Thereafter, Invar pendulum rods are to be found in the best long-case clocks.

Further escapements

There are some mechanisms which ingenious makers will always strive – rightly or wrongly – to improve. The quest for a pendulum swinging as freely as possible resulted in the gravity escapement, in which the pendulum gains its impulse from a weighted arm acted upon by gravity. The force acting on the pendulum – when it does act – is therefore constant; otherwise the pendulum swings freely.

The first practical gravity escapement was designed in 1854 by Edmund Beckett Denison – later Lord Grimthorpe. Denison's double three-legged gravity escapement was designed for the clock in St Stephen's Tower,

part of the Houses of Parliament in London. This clock is often referred to as Big Ben; strictly speaking, that is the name of the bell that strikes the hours.

The anchor recoil escapement invented about 1671. The 'anchor' tilts back and forth as the pendulum swings. As one pallet of the anchor releases one tooth of the escape wheel, another tooth meets the other pallet of the anchor.

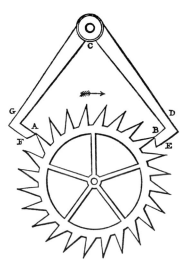

In Graham's dead-beat escapement, the teeth of the escape wheel and the pallets of the anchor are so shaped that there is no recoil.

A repeating mechanism – for striking the hour on demand – attributed to Thomas Tompion. The design ensures that the correct hour is struck even when a step of the snail is in transit.

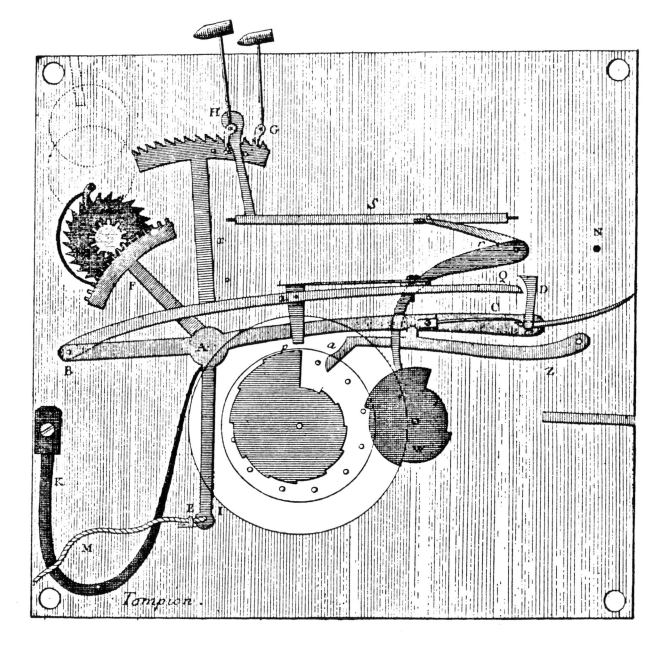

The Biſſextile, or Leap-year.

	Jan. M	Jan. S	Febr. M	Febr. S	Marc M	Marc S	April M	April S	May M	May S	June M	June S	July M	July S	Aug. M	Aug. S	Sept. M	Sept. S	Oct. M	Oct. S	Nov. M	Nov. S	Dec. M	Dec. S
1	8	47	14	49	10	00	0	41	4	10	0	59	4	47	4	26	3	58	13	22	15	19	5	28
2	9	10	14	48	9	43	0	24	4	11	0	47	4	55	4	16	4	19	13	36	15	10	4	59
3	9	32	14	46	9	26	0	8	4	12	0	34	5	2	4	5	4	39	13	49	15	01	4	31
4	9	54	14	43	9	9	0	7	4	13	0	22	5	9	3	54	5	00	14	2	14	50	4	2
5	10	15	14	40	8	51	0	22	4	12	0	10	5	15	3	43	5	20	14	14	14	38	3	33
6	10	36	14	36	8	33	0	37	4	11	0	03	5	20	3	31	5	41	14	25	14	26	3	3
7	10	55	14	31	8	15	0	52	4	10	0	16	5	25	3	18	6	1	14	37	14	13	2	33
8	11	14	14	26	7	57	1	6	4	8	0	29	5	30	3	5	6	22	14	47	14	00	2	3
9	11	32	14	20	7	39	1	19	4	5	0	42	5	34	2	52	6	43	14	57	13	45	1	33
10	11	49	14	13	7	20	1	31	4	2	0	55	5	37	2	38	7	3	15	6	13	30	1	4
11	12	5	14	5	7	1	1	44	3	59	1	7	5	40	2	24	7	24	15	15	13	13	0	34
12	12	22	13	57	6	43	1	57	3	54	1	20	5	43	2	9	7	44	15	24	12	56	0	4
13	12	37	13	48	6	24	2	9	3	50	1	33	5	45	1	54	8	4	15	30	12	38	0	26
14	12	51	13	39	6	05	2	19	3	45	1	46	5	45	1	38	8	24	15	36	12	19	0	56
15	13	5	13	29	5	46	2	30	3	39	1	59	5	46	1	22	8	43	15	42	12	00	1	26
16	13	18	13	18	5	27	2	41	3	33	2	11	5	46	1	5	9	3	15	47	11	40	1	56
17	13	30	13	7	5	9	2	51	3	26	2	23	5	45	0	48	9	23	15	51	11	20	2	25
18	13	41	12	56	4	50	3	0	3	19	2	35	5	44	0	31	9	42	15	54	10	59	2	34
19	13	51	12	44	4	31	3	8	3	11	2	47	5	42	0	13	10	2	15	57	10	37	3	23
20	14	0	12	32	4	13	3	16	3	3	2	59	5	40	0	5	10	21	15	59	10	14	3	52
21	14	9	12	18	3	54	3	24	2	54	3	10	5	37	0	22	10	39	16	00	9	50	4	21
22	14	17	12	5	3	36	3	32	2	46	3	22	5	33	0	40	10	57	16	01	9	26	4	40
23	14	24	11	51	3	17	3	39	2	37	3	33	5	29	0	59	11	15	16	00	9	2	5	16
24	14	30	11	36	2	59	3	45	2	27	3	44	5	25	1	19	11	32	15	59	8	37	5	43
25	14	35	11	21	2	40	3	50	2	17	3	54	5	19	1	39	11	4	15	57	8	11	6	11
26	14	39	11	5	2	22	3	54	2	6	4	4	5	13	1	58	12	6	15	54	7	45	6	37
27	14	43	10	50	2	5	3	58	1	56	4	13	5	07	2	17	12	22	15	50	7	19	7	3
28	14	46	10	34	1	47	4	2	1	45	4	22	5	0	2	37	12	37	15	46	6	52	7	29
29	14	47	10	17	1	30	4	5	1	34	4	31	4	52	2	57	12	53	15	40	6	24	7	54
30	14	49			1	13	4	8	1	22	4	39	4	44	3	18	13	8	15	34	5	57	8	18
31	14	49			0	57			1	11			4	35	3	38			15	27			8	41

By 1675, pendulum clocks were accurate enough to enable John Flamsteed (1646–1719), the first Astronomer Royal, to calculate the *equation of time* – the difference between 'solar time' (as shown by a sundial) and 'mean solar time' as indicated by a mechanical device controlled by an escapement. The idea was so novel that earlier long-case clocks often have a table of differences fastened inside the case. At last they could be used to check the sundial, rather than *vice versa*.

Right Four types of compensated pendulum (left to right): with wooden rod and auxiliary bob for fine regulation; with invar rod and zinc bob with tray for compensating weights; grid-iron; mercury-compensated.

Derek Roberts Antiques, Tonbridge

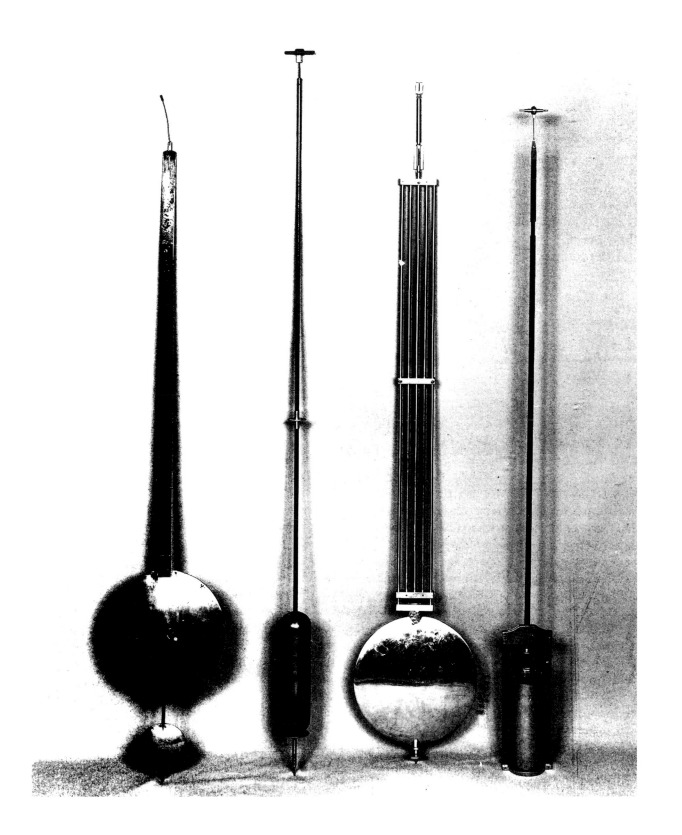

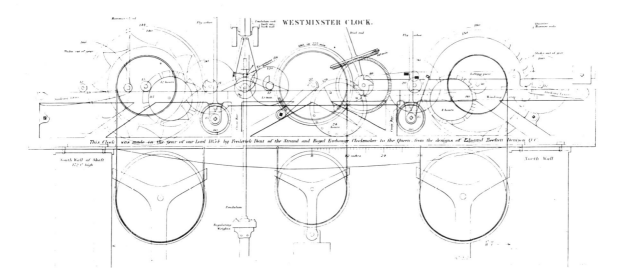

A diagrammatic elevation of the Westminster clock 'made in the year of our Lord 1854 by Frederick Dent of the Strand and Royal Exchange, Clockmaker to the Queen, from the designs of Edmund Beckett Denison, QC.'

Right
Denison's four-legged gravity escapement. The diamond-shape is formed by two separate members hinged at their tops. They are pushed sideways in turn as the pendulum swings to and fro. This movement releases the tip of one of the legs of the escapement but, as it turns, the tip of the opposite leg meets the pallet on the other side. Denison's escapement for the Westminster Clock has a double three-legged escapement, rather than a single four-legged one.

Left
Sir William Congreve's clock of 1808. The timing mechanism is a ball which takes a constant time (in this case 15 seconds) to roll down a zig-zag ramp. When the ball reaches the end of its travel it touches a trigger, the ramp tilts the other way, and the ball returns.

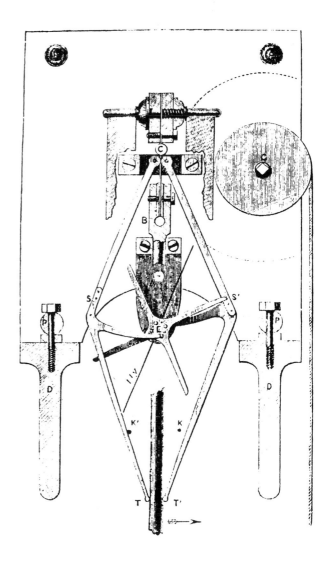

Portable timepieces

Springs and watches

A clock driven by a coiled spring is smaller and neater than one driven by weights, and certainly more portable. A letter from the Italian clockmaker Bartholomew Manfredi to the Marchese di Manta dated November 1462 offers him a 'pocket clock' better than that belonging to the Duke of Modena.

Spring-driven clocks appeared in Italy towards the end of the 15th century; Peter Henlein of Nuremberg, Bavaria, who qualified as a master locksmith in 1509, was building them in 1510. A Henlein watch of iron survives at the Philadelphia Memorial Hall. In 1511, Johannes Cocclaeus of Nuremberg wrote:

From day to day more ingenious discoveries are made; for Petrus Hele [Henlein], a young man, makes things which astonish the most learned mathematicians, for he makes out of a small quantity of iron horologia devised with very many wheels, and these horologia, in any position and without any weight, both indicate and strike for 40 hours, even when they are carried on the breast or in the purse.

Watches appeared in France at about the same time, and in England later in the 16th century. We know of at least two watches by Francis Noway of Brabant, who was working in London between 1580 and 1583. Bartholomew Newseam, Queen Elizabeth I's clockmaker, who flourished between 1568 and his death in 1593, also made watches. The first watch with a second hand – and one of the earliest with a minute hand – was made by John Fitter of Bermondsey (London) about 1665.

The development of the spring enabled timepieces to become smaller and smaller. The 16th-century 'drum clock' was about nine inches in diameter and five inches thick. Drum clocks were often very ornate with, for example, built-in astrolabes, striking action and alarms. Later came the causter watch, two or three inches in diameter and one inch thick. Although such models became more intricate and highly decorated, they were not particularly accurate until 1660, when

Robert Hooke invented the balance wheel and hairspring, made practicable by his friend Thomas Tompion. George Graham of dead beat escapement fame invented his cylinder escapement for watches in 1725.

The first 'portable clock' by Jacob Zech of Prague, 1525 (see page 48).

Right

The earliest known illustration of a watch: a painting by an Italian master about 1560. The watch is a typical mid-16th century German model; on the table are its detachable alarm mechanism and its carrying case.

Science Museum

Right
John Arnold (1736–1799)
of Bodmin in Cornwall
was apprenticed to his
watchmaker father. He
later moved to London,
and in 1764 presented
King George III with a
repeating watch about
three-quarters of an inch
(19 mm) in diameter set in
a ring. This achievement
made him famous, and he
later rejected a request
from the Empress of
Russia for a similar model.
Among other interests, he
produced several marine
chronometers, and
patented 'the modern'
chronometer escapement
(No. 1113 of 1775). In this
portrait by R Davy, he is
seen with a drum clock.

Compensation

One of the problems of using a spring to drive a mechanism is that its force is reduced as it 'runs down'. Two ways of counteracting this were devised: the fusee and the stackfreed. The fusee is a conical drum driven by the spring barrel, which transmits the force to the train. As the force of the spring diminishes, so does the driven diameter of the fusee increase to compensate for the dwindling power. Leonardo da Vinci sketched a fusee (1485–90) but its significance is unclear. The first to use the device in a timekeeping mechanism is thought to be Jacob Zech (or Czech) of Prague in 1525.

The stackfreed is a simpler and less efficient device: a strong spring acts on a snail-shaped cam which turns once as the mainspring runs down. The shape of the snail – and different makers had their pet shapes – is such that the force in the stackfreed spring helps the mainspring. Provision is made to prevent overwinding for, if the end of the stackfreed spring were to jump the snail, the mechanism would become locked.

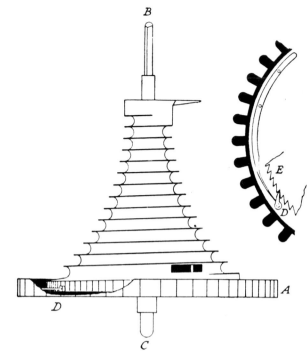

The conical fusee, which is pulled round by the spring barrel to compensate for the diminishing power of the spring.

Frederick Kehlhoff's patent (No. 819 of 1764) showing a stackfreed. Pinion A is carried by the spring barrel, and drives wheel B, which carries snail cam C acted upon by spring D. As the mainspring runs down, the stackfreed spring supplements its power.

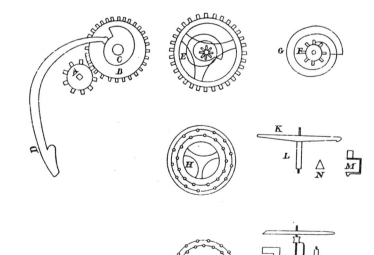

The fusees in this
table clock are
clearly visible.

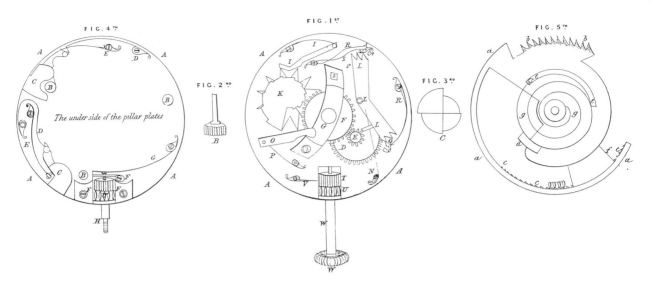

Jewelling

The smaller the mechanism, the worse the effect of friction. Jewelled pivots were introduced to reduce that friction, and were patented in 1704 by the Swiss geometrician and optician Facio de Duillier and the French watchmakers Peter and Jacques Defaubre, all of whom worked in London. Defaubre's first jewelled watch is said to have been worn by Sir Isaac Newton. The art of jewelling was a closely-guarded secret through the 18th century. Thomas Earnshaw (1749–1829) records that, having served a 14-year apprenticeship as a watchmaker, he wished to become a watch jeweller. He was unable to afford the '100 guineas [£105] for the tools and 100 guineas for the sight of them' (payable because seeing them would have revealed the secrets) so he went away and invented his own tools 'so few and simple that the whole cost of them . . . did not exceed four pounds'.

As the art of watchmaking developed, so did the quest for miniaturization. One approach to achieving the levels of production and accuracy demanded as the 18th century progressed was specialization, each part of the watch having its specialist maker.

Patent (No. 1021 of 1772) for a striking watch by Robert Webster of Whitby, Yorkshire.

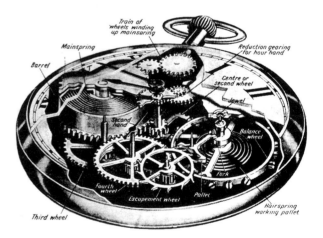

The interior of a watch, showing the escapement. The escape is controlled by the oscillations of the balance wheel, in turn controlled by the coiled hairspring. The action is reminiscent of the verge-and-foliot escapement. The pivots of the escape work are jewelled.

A large silver
coach watch by
William
Knottesford
(1656–93).

Two movements
in one case – a
double watch
with synchronous
balances by the
French maker
Abraham Louis
Breguet
(1747–1823).

'Puritan watch' by Edward East (1610–90), clockmaker to King Charles I.

A French clockmaker of unparalleled ingenuity, Abraham Louis Breguet was always ready to meet customers' whims and demonstrate his capabilities. In this 'Synchronizer', the watch is placed in the cradle on the top at bedtime; in the morning, it has been wound and set right automatically.

Wristwatches

By 1790, Jacquet-Droz and Leschot of Geneva had produced 'a watch to be fixed on a bracelet'. The first surviving appears to be that made by the Parisian jeweller Nitot for the Empress Josephine in 1806. Three years later, the Empress presented her daughter-in-law with a pair of gold bracelets set with pearls and emeralds; one carried a calendar and the other a watch.

The first wristwatches for men – naval officers at sea – were commissioned by the German admiralty in 1880. Men's watches were then manufactured for a wider market, but there appeared to be a demand only in Peru; dealers in the USA would have nothing to do with them. A women's fashion for wearing wristwatches started in Paris in 1908; men considered them

The Waterbury watch (Series E); on its back is inscribed the warning: 'Don't remove this cap unless you are a practical watch repairer.'

Right
The first self-winding watch, patented by John Harwood in 1924 (No 218487). As the wearer moves, the weight A swings between the buffer stops C–C1 and winds the mainspring via a pawl and ratchet.

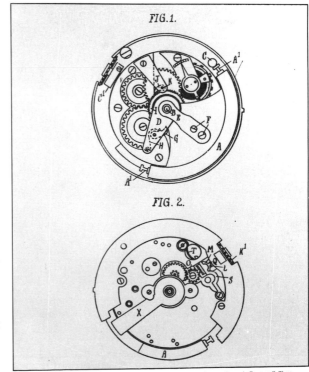

281 The power for winding the watch comes from the vibrations of the weight A between the two buffer stops C. These vibrations are activated by the movements of the wearer and wind up the main spring of the watch by means of a pawl acting on a ratchet wheel.

effeminate until their undoubted convenience was established during the First World War.

Until then, pocket watches had been the norm, perhaps kept in fashion by the opportunity they afforded for sporting a heavy gold chain across the midriff. A watch was a symbol of status; you had to work for it. In Washington, DC, Jason R Hopkins dreamed of producing a watch for 50 cents, and in 1875 built and patented a prototype in which the entire movement rotated once every hour, carrying the minute hand.

THE MARINE CHRONOMETER
Measuring speed

It is important to navigators to know the time accurately, because (until comparatively recently) that was the only way of knowing position. The rate of travel can be measured accurately enough by sand-glass and knotted string (hence the term 'knot' for nautical miles per hour). The method was used by the ancient Greeks, and was developed in the late 16th century. At first, the count was the number of knots seven fathoms (42 feet) apart passing in 30 seconds. A speed of 'one knot' was thus 5040 feet per hour.

However, later surveys adjusted the length of the nautical mile, and the speed in knots became the number of knots four fathoms (24 feet) apart passing in 14 seconds. Some preferred to count knots eight fathoms apart over 28 seconds; 14- and 28-second sand glasses were made for timing. The nautical mile here appears to be about 6171 feet; nowadays the UK knot is 6080 feet per hour, 1.00064 times the International knot (used by the United States since 1954).

The prizes

From 1598, several large prizes were offered for the maker who could produce an accurate, all-weather, ship's chronometer. In 1707, an English fleet was wrecked off the Scillies through a mistake in reckoning. The government set up a Board of Latitude which, in 1714, offered a prize of £20,000 for the long-sought-after marine chronometer.

Huygens submitted a pendulum clock which (not surprisingly) failed the foul-weather test. John Harrison (1693–1776) of London, who had served an apprenticeship as a carpenter, devoted much of his long life to perfecting a marine chronometer. His first weighed about 70lb (32kg) and was fairly successful, but it was not until his No 4 model of 1762 was found to be accurate to 1 minute 54.5 seconds in 147 days (about 0.0009 per cent) that he was awarded the prize; even then, he had to petition King George III for the money.

Without detracting from Harrison's achievement, it is said that it owed as much to the improvements in manufacturing techniques and spring steel as to its maker's abilities.

In 1797, the British Government sought to raise money by imposing an annual clock tax – five shillings for public and domestic clocks, ten shillings for watches carried on the person, and two shillings and sixpence for timekeepers falling outside the first two categories. The difficulties of collecting the tax are obvious, and it caused such a decline in the industry that thousands were put out of work. Fortunately, good sense prevailed; the tax was repealed in 1798, and some compensation paid to the industry. The legacy of the clock tax is the 'Act of Parliament Clock', a large-faced timepiece widely bought and displayed by hoteliers and innkeepers of that time for the benefit of their customers.

Right
Harrison's celebrated Marine Timepiece.

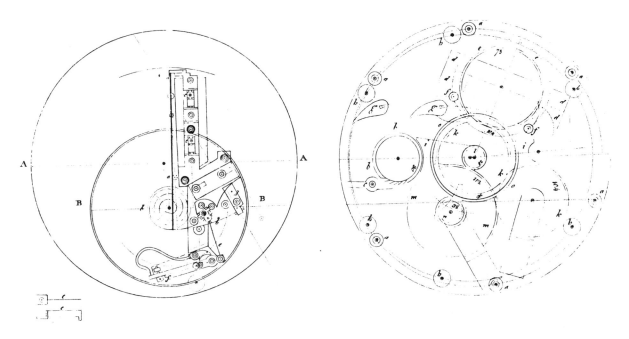

John Harrison's fourth chronometer, which won him the £20,000 prize in 1762. Britten writes: 'It has been stated that the piece hung in gimbals. This is not the case; it reposed on a soft cushion, and on its trial voyages was carefully tended by William Harrison [the maker's son] who avoided position errors as far as possible by shifting the timekeeper to suit the *lie* of the ship.' (Britten)

Nowadays, one's position can be fixed with great accuracy if one has the equipment to receive and interpret satellite signals. Even without such equipment, radio has wiped out the problem of knowing what time it is.

The timetable

At one time, the 'day' was measured as the interval between two appearances of the Sun overhead at 'noon'; indeed, the Sun was used to set mechanical clocks.

However, as mechanical clocks became more and more accurate, it became apparent that 'days' as measured by the Sun in any given place were not all of the same length, and so the 'mean solar day' was introduced.

But this was not all. The time at which the Sun is overhead varies as one travels round the earth longitudinally. If you rely on the Sun, the difference in time between Land's End (the most westerly part of England) and Yarmouth (the most easterly) is about half an hour. Across the United States, the difference is over five hours.

As modern society developed, the need for a standard time grew. Stage-coach operators had long been concerned by the fact that each town and village had its own time, but the pace was leisurely enough to take account of this. The first to do something about the problem were the railway companies who, from the 1830s, worked to a 'railway time' which assumed that time is 'the same' all over the country. Fortunately, the growth of the railway system was accompanied by the development of the telegraph, which made it easier to synchronize clocks throughout the system.

Because of the greater distances, the problem in North America was even more acute, but it was not until 1870 that Charles Ferdinand Dowd proposed a system of national time for railroads, whereby the continent would be divided into time zones, each one hour wide, and within any one of which time would be standard. Dowd's system was introduced in 1883, three years before the opening of the first East–West Canadian Pacific Railroad.

LONDON TO PARIS.					
Dates.	Leave London.	Date.	Leave Folkest'ne Harbour.	Leave Boulogne.	Arrive in Paris.
Wednesday Apr. 1	4 30 p.m.	1	7 30 p.m.	1 0 a.m	9 25 a.m.
Thursday 2	4 30 ...	2	8 0 ...	1 0 ...	9 25 ...
Friday 3	6 30 ...	3	10 0 ...	6 0 ...	1 20 p.m.
Saturday 4	...	...		...	...
Sunday 5	6 30 ...	5	10 0 ...	6 0 ...	1 20 ...
Monday 6	6 30 ...	6	10 30 ...	6 0 ...	1 20 ...
Tuesday 7	8 55 ...	7	Midnight.	6 0 ...	1 20 ...
Wednesday 8	8 55 ...	8	Midnight.	6 0 ...	1 20 ...
Thursday........ 9	8 55 ...	10	1 0 a.m.	6 0 ...	1 20 ...
Friday 10	8 55 ...	11	2 0 ...	6 0 ...	1 20 ...
Saturday 11	...	...		...	...
Sunday 12	8 55 ...	13	4 0 ...	9 0 ...	4 15 ...
Monday 13	8 55 ...	14	5 0 ...	9 0 ...	4 15 ...
Tuesday 14	4 30 ...	14	7 0 ...	1 0 ...	9 25 a.m.
Wednesday 15	4 30 ...	15	7 30 ...	1 0 ...	9 25 ...
Thursday........16	6 30 ...	16	7 50 ...	1 0 ...	9 45 ...
Friday 17	6 30 ...	17	10 0 ...	6 0 ...	1 20 p.m.
Saturday 18	...	...		...	...
Sunday 19	6 30 ...	19	10 0 ...	6 0 ...	1 20 ...
Monday 20	6 30 ...	20	10 30 ...	6 0 ...	1 20 ...
Tuesday 21	8 55 ...	21	Midnight.	6 0 ...	1 20 ...
Wednesday 22	8 55 ...	22	Midnight.	6 0 ...	1 20 ...
Thursday 23	8 55 ...	23	Midnight.	6 0 ...	1 20 ...
Friday 24	8 55 ...	25	1 0 a.m.	6 0 ...	1 20 ...
Saturday 25	...	...		...	...
Sunday 26	8 55 ...	27	3 0 ...	6 0 ...	1 20 ...
Monday 27	8 55 ...	28	4 0 ...	9 0 ...	4 15 ...
Tuesday 28	8 55 ...	29	5 0 ...	9 0 ...	4 15 ...
Wednesday 29	4 30 ...	29	7 15 p.m.	1 0 ...	9 25 a.m
Thursday 30	4 30 ...	30	7 30 ...	1 0 ...	9 25 ...

Part of a timetable for the London–Paris service offered by the South Eastern Railway Company in April 1863. The crossing was from Folkestone to Boulogne, and the journey took just under 19 hours. Today, it takes about 10 hours; the Channel Tunnel halves this.

Standard time

The growth of various sorts of commercial activity led to the examination of systems for introducing world standard times. In 1884, it was internationally agreed that the Greenwich Meridian should be the prime meridian of longitude (*ie* 0°). At that same conference, the universal day (which was to begin at midnight) was defined 'for all purposes for which it may be found convenient, and which shall not interfere with the use of local or other standard time where desirable.' In Great Britain, Greenwich Mean Time became the standard.

The Earth being (by definition) 360° in circumference, 360°= 24 hours, so 15° is therefore equivalent to one hour. The hour-wide time zones imaginarily inscribed on the earth's surface are related to boundaries as convenient – indeed, some are half-an-hour wide. Universal Time (UT) is related to midnight Greenwich Mean Time (GMT). As you move east, standard times become later and later; as you move west, they become earlier.

Although today it would be foolhardy not to accept one's UT, the system was not immediately appreciated and adopted all over the world without question. Some communities were as wedded to their local time as to their language; however, though speech and dialect is now recognized as a precious thing, there can be few living in the Western world who can afford to ignore standards of time.

Time balls

Although we have stressed the importance of time zones on land, knowing the 'exact' time at sea is necessary for fixing one's position. Armed with a sextant and a chronometer set to the time at the home port, the navigator can work out the position of the vessel. As we have seen, chronometers of the necessary accuracy were developed from the end of the 18th century, but how to fix the time at one's home port?

To avoid ships' timekeepers having to go ashore, a visible time signal was needed, and the first was the 'time ball' which, set on high, drops at a known time each day. This device was proposed by Capt Robert Wauchope of the British Royal Navy in 1824. A manually-operated device was first set up at the Royal Observatory, Greenwich, in 1833. In due course, time balls were erected in London and in many ports and, in 1862, time balls in The Strand and Cornhill in London, and at the ports of Deal and Liverpool, were dropped by telegraph signals from Greenwich to inaugurate Greenwich Mean Time. In the United States, the Naval Observatory was set up at Washington in 1845; one of its tasks was to provide a national time service, and it first sent out automatic time signals in 1880.

The telegraph

Advances in telegraphy were spurred by the railways (who needed it for operational reasons) and by the Post Office (who rightly saw it as a way of communication).

The telegraph provided a means for synchronizing clocks over wide areas. At first, signals were sent to mark pre-arranged times, but it was not long before master clocks were set up to transmit the time – as a series of electrical pulses – to slave clocks.

Radio times

As wireless telegraphy developed, it became possible to send out radio time signals, the first of which was transmitted from the Navy Yard at Boston, Massachusetts, on 9 August 1905. The first French time signal was transmitted from the Eiffel Tower in Paris on 25 June 1910. When the British Broadcasting Company was set up in 1922, it took its time from Paris, but was soon connected to the Royal Observatory at Greenwich. The British tradition of ushering in the new year to the sound of Big Ben began as 1923 gave way to 1924, the year in which the 'pips' were first heard.

There is a story that the Director of the Paris Observatory was making an official visit to the Eiffel Tower, and enquired about the time signal transmissions: 'Tell me, how do you set your clocks?'
'Why, M le Directeur, we telephone your Observatory twice a day and check our clocks against yours. But I've often wondered . . . how do you maintain *your* time standards?' After some reddening discussions, it emerged that the Observatory set its clocks by listening to the time signals from the Eiffel tower.

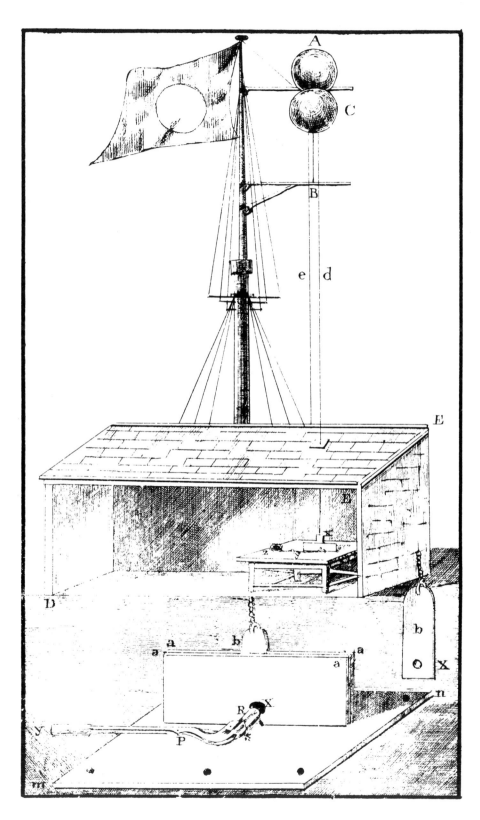

Admiral Robert Wauchope's 'plan for ascertaining the rates of chronometers by an instantaneous signal.' The balls are five feet (1.5 m) in diameter, made of black canvas on a wire frame. The top ball is fixed. The lower ball is released by someone watching a clock and pulling a lever four-fifths of a second before the agreed time (!) This is because zero hour is taken as the time when the balls are separated by their own diameter, and this takes four-fifths of a second after release.

Electrons and atoms

Electricity harnessed

It is not surprising that the 'mysterious force of electricity' should have been applied to clocks. At first, 'static' electricity – the sort you get from a dry cat – was used to ring a bell by attraction and repulsion of a pith ball suspended from a silken thread.

The first electric cell – which produces an electric current – was built by Alessandro Volta in 1800, and in the 1820s André Marie Ampère discovered that a current flowing through a coil turned that coil into a magnet. Here was a gift to inventors, though it was not until 1841 that the 'Father of Electric Horology' – Alexander Bain (1810–77) – took out a comprehensive patent for an electrically-driven clock. The far-sighted Bain covered most of the applications, including electromagnetic drive, and using a master clock to control a number of slave clocks.

The following year, Dr Matthäus Hipp of Neuchatel, Switzerland, invented the free pendulum system which applies a maintaining impulse only when the swing of the pendulum falls short of a predetermined arc. However, the impulse is not applied at regular intervals, so the device cannot be used to drive a slave clock, but the principle was developed and became the basis of successful systems which continue to this day.

The electrical pioneer SZ Ferranti built the first alternating current (AC) generator at Deptford (London) in 1895. It soon became apparent that an AC supply of

One of the numerous drawings from Alexander Bain's patent (No. 8783 of 1841), in which he described almost every aspect of electrical timekeeping. The diagram shows the coils of an electromagnet whose pole-pieces *aa* attract an armature *b*, which action pulls the ratchet wheel round via hooked arm *c*.

controlled frequency could be used to drive a simple 'synchronous' clock mechanism. As the AC supply became the norm, and several connected generators were needed to supply the demand for power, it was essential to have a standard frequency. In America, where the frequency standard was 60Hz, HE Warren introduced the synchronous clock in 1916. By World War II, more than half the clocks sold in the USA were of the synchronous type.

In England, the Grid System for electrical distribution was established in 1925, with a standard frequency of 50Hz. Soon, the synchronous clock took over from all other types for everyday use in home and office.

The ultimate in pendulum timekeeping was the Shortt free-pendulum clock, installed at the Edinburgh Observatory in 1921. This comprised two pendulums, one of which (the slave) drove the mechanism while the other (the master) swung freely except for a maintaining impulse every half minute. Accurate to a fraction of a second per year, the Shortt clock was soon to be found in astronomical observatories throughout the world.

Pond's winder

The perfectly reliable mechanical clock – especially the turret clock with heavy weights – can be enhanced by an electrically-assisted winding mechanism. A constantly-wound spring provides constant power.

In 1881, Chester H Pond of Brooklyn, NY, invented a means of winding a spring clock at hourly intervals, thus improving the accuracy of the clock.

The quartz clock

A quartz crystal exhibits the 'piezoelectric effect' – that is, it will produce electricity if distorted. Conversely, if the crystal is subjected to an electric charge, it will vibrate. The frequency at which it vibrates is determined by its dimensions, and crystals can be cut and finished to vibrate at a desired frequency with extreme accuracy – particularly when kept in an enclosure at a controlled temperature. Such a crystal may be used to control the frequency of vibration of an electrical circuit (typically at many kilohertz). This is used – directly or indirectly – to control a timekeeping mechanism.

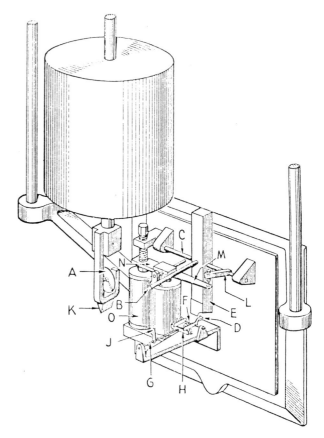

The 'inertia escapement' of the first Shortt free-pendulum clock, arranged to deliver a impulse below the bob on each swing. When the pendulum passes through its central position, K triggers J, releasing rod B, which imparts an impulse to wheel A. Contacts L and M close, and the electromagnet O returns bar B to its original position.

The principle was described by the Americans HE Horton and WA Marrison in 1928, and was put into practice in the early 1940s. Although it was not realized at the time, the quartz movement was to revolutionize the world's timekeeping industries.

The quartz movement may be accurate to 0.0001 second per year or better. With the development of microelectronics, the quartz frequency standard was applied to wristwatches in 1969. Later, such watches were also provided with functions such as alarms and stop timing, and even built-in calculators. The limit of miniaturization is set by the keenness of the human eye, and the size of the control buttons which can be pressed.

The atomic clock

The idea of using atomic or molecular vibrations to control the frequency of an electronic oscillator was put forward in 1946.

The first 'atomic' clock was built in 1948 by Harold Lyons. It used the vibrations of ammonia molecules to provide a standard ten times more accurate than the quartz crystal. Molecules of ammonia are injected into a cavity between metal electrodes charged to a high voltage. The ammonia molecule is not symmetrical and, according to whether a given molecule has a high or a low energy content, so it aligns itself one way or another in the electric field.

By appropriate manipulation of the charges, the high-energy molecules can be separated from their fellows and fed into a cavity resonator, where they control the frequency of vibration of an electrical signal with very high accuracy. The signal may then be treated as in the quartz clock, and made to drive a system for measuring the passage of time.

The National Bureau of Standards, Washington, DC, adopted the ammonia clock standard accurate to one part in 100,000,000 – in 1947.

The caesium frequency standard was devised in 1955 at the National Physical Laboratory (NPL), Teddington, England. Over the next three years, NPL and the US Naval Laboratories carried out joint experiments to produce a commercially viable standard of time of hitherto undreamed-of accuracy.

In this standard, caesium atoms swing between two states of energy in a resonant cavity, controlling the circuit that makes the cavity vibrate. Once the device has been brought to its working frequency, it stays there, keeping time to one second in 300,000 years – better than the Earth, which may vary by one millisecond each day.

This is the principle of the maser – which does for microwave radiation what the laser does for light, providing a source of fixed frequency by harnessing the stability of molecular vibration.

The twin atomic hydrogen masers developed at Harvard University and installed at the US Naval Research Laboratory, Washington, DC, in 1964 were accurate to within one second in 1,700,000 years; unfortunately, 'the uncertainty of the fundamental frequency was greater than the stability of the clock'.

The caesium standard was, after all, more accurate, and in 1967 the second was redefined as '9,192,631,770 periods of radiation emitted or absorbed in the hyperfine transition of the Cs133 atom'.

WHICH TIME?

Today, the Internationale Bureau des Poids et Mesures – (BIPM) at Sèvres, near Paris, coordinates worldwide time measurement, which is based on the International Atomic Time Scale (TAI).

In 1956, three sorts of universal time were recognized:
   UT_0 is mean solar time obtained by direct astronomical observation.
   UT_1 is UT_0 corrected for the effects of the wanderings of the position of the earth's pole.
   UT_2 is UT_1 corrected for seasonal fluctuations in the rate of rotation of the earth.
By international agreement, time is now regulated by coordinated universal time (UTC) kept as close as possible to UT_1. It runs as fast as TAI.

In 1972, UTC was set 10 seconds slower than TAI. Now UTC and UT_1 are regularly compared, and as soon as the two start to deviate by more than 0.9 seconds, a 'leap second' is added to UTC. Because of the addition of leap seconds UTC always differs from TAI by an integral number of seconds.

Between 1972 and 1990, 15 seconds were added, which shows that the rotation of the earth slowed by an average of 0.83 seconds per year.

Two atomic clocks at the National Physical Laboratory, Teddington, England. The vertical tube in the foreground is the main part of the standard clock; the trolley on the left is an experimental clock. NPL now transmits a radio signal for use by self-regulating clocks.

National Physical Laboratory

Chronology

3761	Start of the Hebrew calendar	1505	First public dial
753	Start of the Roman calendar	1581	Galileo's pendulum
325	Council of Nicea moves toward modern calendar	1582	The Gregorian calendar
610	Chinese steelyard clepsydra	1656	Huygens' pendulum
622	Start of the Muslim calendar	1676	Barlow's striking mechanism
1088	Chinese water escapement	1704	Jewelled pivots
1283	Dunstable Priory clock	1762	Harrison's prize chronometer
1325	Norwich Cathedral clock	1790	Wrist watch
1327	St Albans clock	1824	Proposal for a time ball
1340	Earliest known picture of sandglass	1841	Bain's electric embodiments
1348	de'Dondi's astronomical movement	1895	AC electricity supply introduced
1386	Salisbury Cathedral clock	1905	Radio time signals
1392	Wells Cathedral clock	1921	Shortt free-pendulum clock
1462	Earliest mention of a 'pocket clock'	1928	Quartz crystal oscillator
		1955	Caesium clock

Further reading

Magdalen Bear
Days, Months & Years
Stradbroke, Tarquin Publications 1989

FJ Britten
Old Clocks and Watches and their Makers
various editions

Eric Bruton
History of Clocks and Watches
New York, Outlet 1989

Ward L Goodrich
The Modern Clock: A Study of Time Keeping Mechanism, Its Construction, Regulation and Repair
Arlington Books, 1985

Cedric Jagger
The World's Great Clocks and Watches
London, Hamlyn 1977

Gerald Jenkins & Magdalen Bear
Sundials & Timedials
Stradbroke, Tarquin Publications 1987

James Jespersen & Jane Fitz-Randolph
From Sundials to Atomic Clocks: Understanding Time and Frequency
New York, Dover 1983

David F Landes
Revolution in Time: Clocks and the Making of the Modern World
Cambridge, Harvard University Press 1883

Frank Parise (ed)
Book of Calendars
New York, Facts on File, 1982

VV Psybulsky
Calendars of Middle East Countries
Moscow 1977

Charles Singer *et al*
A History of Technology
Oxford University Press

Hugh Tait
Clocks and Watches
London, British Museum 1990

Leonard Weiss
Watch-making in England 1760–1820
London, Robert Hale 1982

Index